T0137054

Lecture Notes in Networks and Systems 876

The series "Lecture Notes in Networks and Systems" publishes the latest developments in Networks and Systems—quickly, informally and with high quality. Original research reported in proceedings and post-proceedings represents the core of LNNS.

Volumes published in LNNS embrace all aspects and subfields of, as well as new challenges in, Networks and Systems.

The series contains proceedings and edited volumes in systems and networks, spanning the areas of Cyber-Physical Systems, Autonomous Systems, Sensor Networks, Control Systems, Energy Systems, Automotive Systems, Biological Systems, Vehicular Networking and Connected Vehicles, Aerospace Systems, Automation, Manufacturing, Smart Grids, Nonlinear Systems, Power Systems, Robotics, Social Systems, Economic Systems and other. Of particular value to both the contributors and the readership are the short publication timeframe and the world-wide distribution and exposure which enable both a wide and rapid dissemination of research output.

The series covers the theory, applications, and perspectives on the state of the art and future developments relevant to systems and networks, decision making, control, complex processes and related areas, as embedded in the fields of interdisciplinary and applied sciences, engineering, computer science, physics, economics, social, and life sciences, as well as the paradigms and methodologies behind them.

Indexed by SCOPUS, INSPEC, WTI Frankfurt eG, zbMATH, SCImago.

All books published in the series are submitted for consideration in Web of Science.

For proposals from Asia please contact Aninda Bose (aninda.bose@springer.com).

Mostafa Al-Emran · Jaber H. Ali · Marco Valeri ·
Alhamzah Alnoor · Zaid Alaa Hussien
Editors

Beyond Reality: Navigating the Power of Metaverse and Its Applications

Proceedings of 3rd International Multi-Disciplinary Conference - Theme: Integrated Sciences and Technologies (IMDC-IST 2024) Volume 2

Editors
Mostafa Al-Emran
The British University in Dubai
Dubai, United Arab Emirates

Jaber H. Ali
Southern Technical University
Basrah, Iraq

Marco Valeri
Faculty of Economics
University Niccolò Cusano
Rome, Italy

Alhamzah Alnoor
Southern Technical University
Basrah, Iraq

Zaid Alaa Hussien
Southern Technical University
Basrah, Iraq

ISSN 2367-3370 ISSN 2367-3389 (electronic)
Lecture Notes in Networks and Systems
ISBN 978-3-031-51299-5 ISBN 978-3-031-51300-8 (eBook)
https://doi.org/10.1007/978-3-031-51300-8

This Springer imprint is published by the registered company Springer Nature Switzerland AG
The registered company address is: Gewerbestrasse 11, 6330 Cham, Switzerland

Paper in this product is recyclable.

Preface

The Metaverse refers to a three-dimensional virtual space involving social connections. Its management applications landscape is undergoing significant changes not only as a result of global restrictions related to COVID-19 but also due to disruptive innovations in the digital realm. Metaverse has changed people's perception of virtual experiences for management applications. Many practitioners focus on creating virtual experiences and products for consumers. The 3rd International Multi-Disciplinary Conference Theme: Integrated Sciences and Technologies (IMDC-IST 2024) is held to address the theme "Beyond reality: Navigating the power of Metaverse and its applications". The IMDC-IST 2024 brings together a wide range of researchers from different disciplines. It seeks to promote, encourage, and recognize excellence in scientific research related to Metaverse and virtual reality applications. The main aim of the IMDC-IST 2024 is to provide a forum for academics, researchers, and developers from academia and industry to share and exchange their latest research contributions and identify practical implications of emerging technologies to advance the wheel of these solutions for global impact. In line with the Fourth Industrial Revolution goals and its impact on sustainable development, IMDC-IST 2024 is devoted to increasing the understanding and impact of the Metaverse on individuals, organizations, and societies and how Metaverse applications have recently reshaped these entities.

The IMDC-IST 2024 attracted 65 submissions from different countries worldwide. Out of the 65 submissions, we accepted 29, representing an acceptance rate of 44.61%. About 14 papers are accepted in Volume 2. The chapters of Volume 2 explore the diverse implications of the Metaverse in various professional and social sectors. It opens with insights from travel agency employees on Metaverse's impact on their industry and discusses the broader transition of tourism from the physical world to the virtual realm. It scrutinizes the potential impacts of the Metaverse on workplace dynamics, including both opportunities and challenges. This volume also assesses the strategic potential of Metaverse in securing a sustainable competitive edge for businesses and investigates how the Metaverse might enhance or redefine productivity through real-time, multisensory social interactions. Education continues to be a key theme, with discussions on the factors driving school teachers to adopt virtual educational resources to elevate the educational process. Furthermore, the chapters delve into the evolving landscape of digital marketing within the Metaverse, considering brand activity, consumer behavior, and even the influence of religion on marketing strategies in virtual environments. The book addresses serious considerations of social presence, support theories, and privacy risks in adopting Metaverse platforms. Leadership practices within higher education institutions are viewed through the lens of virtual reality, emphasizing the potential for authentic leadership to foster innovative work behavior. The financial sector is not left out, with a case study on virtual reality technology's role in information disclosure within banks. Finally, the volume looks at how marketing mood management within virtual platforms can contribute to organizational sustainability, underscoring the comprehensive impact of the Metaverse across various facets of modern life. Each chapter offers a distinct perspective, providing readers with a well-rounded understanding of the challenges and prospects of the Metaverse.

Each submission is reviewed by at least two reviewers, who are considered experts in the related submitted paper. The evaluation criteria include several issues: correctness, originality, technical strength, significance, presentation quality, interest, and relevance to the conference scope. The conference proceedings are published in *Lecture Notes in Networks and Systems Series* by Springer, which has a high SJR impact. We acknowledge all those who contributed to the success of IMDC-IST 2024. We would also like to thank the reviewers for their valuable feedback and suggestions. Without them, it was impossible to maintain the high quality and success of IMDC-IST 2024.

Mostafa Al-Emran
Jaber H. Ali
Marco Valeri
Alhamzah Alnoor
Zaid Alaa Hussien

Organization

Conference General Chairs

Mostafa Al-Emran	The British University in Dubai, Dubai, UAE
Alhamzah Alnoor	Southern Technical University, Basrah, Iraq

Honorary Conference Chair

Rabee Hashem Al-Abbasi	Chancellor of Southern Technical University, Basrah, Iraq

Conference Organizing Chair

Jaber H. Ali	Dean of Management Technical College, Southern Technical University, Basrah, Iraq

Program Committee Chair

Marco Valeri	Faculty of Economics, Niccolo' Cusano University in Rome, Italy

Publication Committee Chairs

Mostafa Al-Emran	The British University in Dubai, Dubai, UAE
Jaber H. Ali	Dean of Management Technical College, Southern Technical University, Basrah, Iraq
Marco Valeri	Niccolo' Cusano University, Italy
Alhamzah Alnoor	Southern Technical University, Basrah, Iraq
Zaid Alaa Hussein	Southern Technical University, Basrah, Iraq

Conference Tracks Chairs

Sammar Abbas	Institute of Business Studies, Kohat University of Science and Technology, Pakistan
Gül Erkol Bayram	School of Tourism and Hospitality Management Department of Tour Guiding, Sinop University, Sinop, Turkey
Marcos Ferasso	Economics and Business Sciences Department, Universidade Autónoma de Lisboa, 1169-023 Lisboa, Portugal
Hussam Al Halbusi	Department of Management at Ahmed Bin Mohammad Military College, Doha, Qatar
Khai Wah Khaw	School of Management, Universiti Sains Malaysia, 11800, Pulau Pinang, Malaysia
Gadaf Rexhepi	Southeast European University, Tetovo, The Republic of Macedonia

Members of Scientific Committee

Hashem Nayef Hashem	Sothern Technical University, Technical College of Management, Basra, Iraq
Gul Erkol Bayram	Sinop University, School of Tourism and Hospitality Management Department of Tour Guiding, Turkey
Akram Mohsen Al-Yasiri	University of Karbala, Iraq
Hussan Al Halbusi	Department of Management, Ahmed Bin Mohammed Military College, Qatar
Muslim Allawi Al-Shibli	Maqal University, Iraq
Sammar Abbas	Institute of Business Studies, Kohat University of Science and Technology, Pakistan
Taher Mohsen Mansour	Shatt Al-Arab University, Iraq
Gadaf Rexhepi	Southern European University, The Republic of Macedonia
Abdul Hussein Tawfiq Shalabi	College of Administration and Economics, University of Basra, Iraq
Marcos Ferasso	Economic and Business Sciences Department, Universidade Autonoma de Lisboa, Portugal
Safaa Muhammad Hadi	Sothern Technical University, Technical College of Management, Basra, Iraq
Nabil Jaafar Al-Marsoumi	Al-Maqal University, Iraq
Muhammad Helou Daoud	Ministry of Higher Education, Iraq

Khai Wah Khaw	School of Management, Universiti Sains Malaysia, Malaysia
Hadi Abdel Wahab	College of Administration and Economics, University of Basra, Iraq
Salma Abdel Baqi	College of Information Technology, University of Basra, Iraq
Imad Abdel Sattar Salman	Southern Technical University, Basrah, Iraq

Publicity and Public Relations Committee

Zaid Alaa Hussein	Southern Technical University, Basrah, Iraq
Alhamzah Alnoor	Southern Technical University, Basrah, Iraq

Finance Chair

Jaber H. Ali	Dean of Management Technical College, Southern Technical University, Basrah, Iraq

Contents

Evaluation of the Metaverse: Perspectives of Travel Agency Employees 1
Beste Demir, Selda Guven, and Bayram Sahin

Transitioning the Tourism Industry from the Real World to the Metaverse 21
Neha Sharma, Neetima Aggarwal, Mahesh Uniyal, Gul Erkol Bayram, and Vijay Prakash

The Impact of Management Information Systems on International Human Resource Management: Moderating Role of Metaverse Culture 37
Hiba Yousif Al-Musawi and Marcos Ferasso

Impact of Metaverse at Workplace: Opportunity and Challenges 54
Bushra Al Harthy, Aseela Al Harthi, Arash Arianpoor, and Ali Shakir Zaidan

Measuring the Possibility of Adopting Metaverse Technology as an Appropriate Strategy to Achieve a Sustainable Competitive Advantage ... 69
Hashim Nayef Hashim Al-Hachim and Adnan Saad Tuama Al-Sukaini

Unveiling the Quality Perception of Productivity from the Senses of Real-Time Multisensory Social Interactions Strategies in Metaverse 83
Abbas Gatea Atiyah

Factors Influencing School Teachers' Intention to Adopt Open Virtual Educational Resources Platform in Saudi Arabia 94
Waleed Saud Alshammri, Siti Mastura Baharudin, and Azidah Bt Abu Ziden

How Are Brand Activity and Purchase Behavior Affected by Digital Marketing in the Metaverse Universe? 112
Nadia Atiyah Atshan, Hasan Oudah Abdullah, Hadi AL-Abrrow, and Sammar Abbas

The Effect of Religion on Metaverse Marketing 129
Bakhtiar Aubaid Sharif

Understanding Metaverse Adoption Strategy from Perspective of Social Presence and Support Theories: The Moderating Role of Privacy Risks 144
Abbas Gatea Atiyah, Mushtaq Alhasnawi, and Muthana Faaeq Almasoodi

Employing Metaverse Technologies to Improve the Quality of the Educational Process 159
Abdulridha Nasser Mohsin, Munaf abdulkadim Mohammed, and Marwa Al-Maatoq

Influence of Authentic Leadership Practices on Innovative Work Behaviour in Higher Educational Institutions: A Virtual Reality Perspective 175
Hafiza Saadia Sharif, Al-Amin Bin Mydin, and Hussain A. Younis

The Role of Virtual Reality Technology in Disclosing Future Information: Evidence from Iraqi Banks 188
Azhaar Al-Ali and Assmaa Mahdi Al-hashimi

The Effect of Marketing Mood Management in Enhancing Sustainability: Evidence from Virtual Marketing Platforms 200
Mortada Mohsen Taher Al-Taie, Ilham Nazem Abdel-Hadi, and Hossam Hussein Shiaa

Author Index 211

Evaluation of the Metaverse: Perspectives of Travel Agency Employees

Beste Demir[1](✉), Selda Guven[2], and Bayram Sahin[1]

[1] Balıkesir University, Balıkesir, Turkey
bestedemir2014@gmail.com, bsahin@balikesir.edu.tr
[2] Çanakkale Onsekiz Mart University, Ezine, Çanakkale, Turkey
seldaegilmezgil@comu.edu.tr

Abstract. The metaverse constitutes a notable facet of the digital transformation underway in the tourism sector. The interplay of this process and consumer activity substantially impacts the global disposition and practices of industry personnel. Although much of the attention centres on the potential changes to the tourism industry's services and customer inclinations engendered by the metaverse, the stakeholder position on this concept is seldom acknowledged. Within the context of this study, the objective is to determine travel agency employees' views on the metaverse and its potential impact on the sector. In the study, which utilised descriptive and content analysis through qualitative research methods, it was found that employees have limited understanding of the metaverse. They acknowledge that there may be challenges with its implementation due to technological and financial obstacles, and they view the metaverse as being most advantageous for marketing purposes. Considering the progression of technology and the speed with which businesses have adapted, it has been determined that travel agencies will be significantly impacted by these innovations. The metaverse and digital transformation are expected to provide striking opportunities, particularly in the field of marketing for travel agencies. Although it is uncertain what changes the technology will bring in the future, it should not be forgotten that failing to keep up with technology can have catastrophic consequences for travel agencies.

Keywords: Metaverse · Meta-tourism · Travel agencies

1 Introduction

The transformative information technologies (He et al. 2018) have become one of the most significant tools of modern times, reshaping practically every aspect of life. The recent crises have increased the demand for virtual travel and tourism experiences (Gursoy et al. 2022; Manhal et al. 2023). Subjected to numerous global changes - technological, political, demographic, and social (Guduraš 2014; Alnoor et al. 2022) - in the tourism industry, a range of opportunities and risks lead to the emergence of new paradigms. The competitive edge of tourism organisations and destinations is highlighted, and the industry is supported by globalisation (Buhalis and O'Connor 2005). Discovering, understanding, and purchasing tourism services have undergone significant changes as a result of the latest advancements in information technologies (Cranmer et al. 2020).

M. Al-Emran et al. (Eds.): IMDC-IST 2024, LNNS 876, pp. 1–20, 2023.
https://doi.org/10.1007/978-3-031-51300-8_1

Virtual, augmented, mixed reality and metaverse can all describe changes in the tourism industry. In this context, virtual reality is a type of human-computer interface that is partially controlled by the user (Helsel 1992) and simulates a real environment (Zheng et al. 1998; Yuviler-Gavish, et al. 2013). Virtual reality is increasingly used in various industries, including tourism (Kim and Hall 2019). Using tools that provide a sensory experience in sight, hearing, touch, and smell, the user can engage with the environment as if it were part of the tangible world (Coelho et al. 2006). Virtual reality enables users to feel immersed in the activity (Pantelidis 1993), while the principal aspect is to give the user the sensation of being in another place (Hoffman 2004).

Applications of augmented reality, which are computer-generated and predominantly supported by graphical content (Han et al. 2014), can be defined as the technology that enables the display of the real environment in the real world (Saragih and Suyoto, 2020) typically require travelling to the destination (Lu et al. 2021). Examples of mixed reality include integrating computer graphics into a real, three-dimensional scene or transforming physical objects into a virtual setting. The initial scenario is often designated as augmented reality, while the latter is termed augmented virtuality (Pan et al. 2006). The outcome of amalgamating the physical and digital worlds is mixed reality, a fusion of augmented and virtual reality (Morimoto et al. 2022). With the use of extended reality, which includes virtual, augmented, and mixed reality, technology can produce entirely or partly synthetic digital environments enabling people to interact (Mystakidis 2022; Aymen et al. 2019). Virtual reality, augmented reality, and mixed reality are only a few emerging technologies that make up the Metaverse (Trunfio and Rossi 2022). The Metaverse offers a service with deeper content and social meaning, unlike virtual and augmented reality which focuses on the physical approach and processing. Additionally, although augmented reality and virtual reality may be used in the Metaverse, they do not always do so (Muhsen et al. 2023; Al-Hchaimi et al. 2023; Chew et al. 2023). Although the platform lacks support for these reality types, it has the potential of becoming a metaverse application. Ultimately, an environment that can accommodate a significant number of people is vital for the Metaverse's social relevance (Park and Kim 2022).

In the gaming industry, metaverses such as Roblox (www.roblox.com), Sandbox (www.sandbox.game), and Fortnite (epicgames.com/fortnite) are increasingly prevalent. The recent announcement by Marc Zuckerberg that his company would change its name to Meta has generated curiosity among a wider audience (Narin 2021). Indeed, according to the 2022 report by McKinsey & Company titled "Value creation in the metaverse," the metaverse is projected to generate a value of £5 trillion by 2030. Additionally, sectors that were early adopters of the metaverse aim to allocate a significant proportion of their digital investment budgets towards its development. Tourism is among the identified sectors, and virtual travel is deemed one of the top five phenomena that customers aspire to witness, based on the 2022 report by McKinsey & Company (Abbas et al. 2023; Bozanic et al. 2023).

Applications that restrict communication and mobility during the pandemic are one of the key factors leading to increased interest in the metaverse. In fact, Kye et al. (2021) suggest that Covid-19 restrictions are mainly responsible for the rapid growth of the metaverse in various fields, including education, healthcare, fashion, and tourism. According to Zhong et al. (2021), the pandemic may lead to a completely new phase of virtual tourist activities in the tourism industry. As noted by Mohanty et al. (2020), promoting virtual experiences is crucial for developing a sustainable tourism model, considering the pandemic's negative impact on mobility levels for tourists. Donthu (2020) argued that the post-pandemic era will give rise to a new world, one that witnesses the unprecedented emergence of online communication, shopping, and entertainment. According to Shubtsova et al. (2020), there is potential for new and innovative business models to replace old models in the tourism industry.

As mentioned previously, the world is undergoing a significant transformation. This adaptation is particularly highlighted. Major global crises, technological progress, and global trends have a significant impact on progress. In terms of sustainability and competitiveness, companies must adapt to these changing circumstances. Being outdated, not keeping up with evolving technologies, and failing to adapt to changing consumer behaviours can have disastrous consequences for tourism companies. According to Ercan (2022), it is crucial to identify the metaverses that will affect tourism, understand how they will manifest, and anticipate the kind of changes that are likely to happen. Learning the opinions of executives in the travel and tourism industry - the key players - on the metaverse is now crucial. In this research, it is aimed to determine how employees of travel agencies interpret the metaverse and its possible implications (Atiyah 2020). As a key proponent of the metaverse, it is anticipated that it will aid in gathering the perspectives of travel agency employees on the metaverse, propagating the metaverse's usage, identifying potential issues and implementing preventative measures.

2 The Metaverse Concept

Based on etymology, the Latin roots of the term "verse" literally translate to "in chorus," meaning all as one and as a whole. The Greek prefix "meta" typically signifies beyond or above. In this regard, the metaverse may contain related and overlapping content. It can refer to something beyond the confines of temporal and spatially defined physical reality, potentially denoting a universe outside the physical realm. It is possible that one or more hypothetical universes exist beyond the one humans currently inhabit (Dolata and Schwabe 2023; Atiyah,2023). Neal Stephenson's 1992 novel Snow Crash popularised the notion of the "Metaverse" (Gursoy et al. 2022), which took on a commercial identity as Second Life, a social virtual world game, in 2003 (Um et al. 2022). According to Dwivedi et al. (2002), the term "metaverse" refers to a shared online space where individuals collaborate to create value using intelligent virtual environments that generate physical products and services.

The real-world metaverse can be differentiated from the virtual-based metaverse depending on the world that is given more prominence. The metaverse demonstrates the idea of integration between the real and virtual worlds. A reality-based metaverse concentrates on enhancing the real world through virtual technology (Ane et al. 2019; Alnoor et al. 2020). As per the definition provided by Lee et al. (2021), the metaverse represents a virtual environment where physical and digital components coexist, enabled by the merger of extended reality and internet and web technologies. Conversely, a metaverse that operates solely virtually offers a novel encounter occurring within virtual surroundings, distinct from the physical realm. Users can utilise augmented or virtual reality to access all experiences and material available on the metaverse from any location, blurring the distinction between the physical and virtual domains. As users reside in both environments simultaneously, thanks to their digital avatars, their actions on either side impact the other in real time (Yemenici 2022).

When examining the early formations of the metaverse, it is evident that network infrastructure based on 2G, 3G, and 4G is provided, while 2D images are produced using mobile phones and mouse/keyboard controls (Jung and Jeon 2022; Atshan et al. 2022). Then, it is evident that the avatars acting as representatives of users in the metadata warehouse, virtual reality, augmented reality, social media, virtual economy, and virtual reality trade support the three-dimensional virtual worlds accessed through specialized devices (Ramadhan et al., 2023; Gatea 2016). Davis et al. (2009) state the metaverse components consist of the metaverse itself, human/avatars, metaverse technology capabilities, behaviours, and outputs. In the metaverse, behaviour arises from avatar interaction and conversation. The outcomes include sub-dimensions such as support from others, perception of quality, self-image, and fear of reconnection. Wang et al. (2021) subdivided the metaverse into five components. These encompass network infrastructure, cyber-reality interfaces, data methods and applications, authentication mechanisms, and content creation. According to Park and Kim (2022), the three key elements of a considerable metaverse are the development of hardware (including GPU memory and 5G), recognition and expression models that leverage hardware parallelism, and the accessibility of intriguing and participatory content.

Representative businesses, typical products, and their development strategies also change due to the various policies of the Metaverse's multiple nations. For example, the United States, the Metaverse's forerunner, has a sizable Metaverse layout that is used in various contexts, including business, gaming, the arts, and social affairs. China has a considerable market, along with successful internet enterprises and applications. While the system in South Korea is administered by the government, in Japan the focus is on applications in animation and video games. German and Italian luxury companies are endeavoring to expand their market share through virtual goods (Ning et al. 2023). Table 1 presents a list of companies that are investing in the metaverse in various countries.

(Ning et al. 2023).

Despite the recognition of Metaverse as the next-generation internet paradigm that allows users to engage in virtual world play, work, and communication, the internet, a fundamental piece of infrastructure in many countries, is wholly or partially unusable (Dahan et al. 2022). Security and privacy pose additional challenges for metaverse enterprises, developers, and users given the potential for privacy violations, identity theft, and fraud. For example, numerous personal assets within the metaverse, such as digital possessions, virtual item identities, cryptocurrency transaction records, and other confidential user data, are lacking protection (Yang et al. 2022). Ethical issues related to the emergence of technology were brought to light by Kwok and Koh (2021). These include concerns over cybersecurity, the right to privacy, as well as negative effects on cognitive and behavioural aspects like technology addiction, antisocial behaviour, and illusory experiences. At this stage, objective assessments of the metaverse topic, which has received some criticism, are thought to be useful in directing developers and key users.

Table 1. Enterprises Investing in Metaverse in Different Countries.

USA	Amazon, Roblox, Facebook, Epic Games, Disney, Snapchat, Nvidia, Microsoft, Dcentraland
China	Tencent, Alibaba, ByteDance, NetEase, Shenzhen Zqgame Co.Ltd, Wondershare Technology Group Co.,Ltd
Japan	Sony-Hassilas, GREE, Avex Business Development-Digital Motion
South Korea	SAMSUNG, SK Telecom, Urbanbase, Metaverse Alliance
England	Sotheby's, Maze Theory
United Arab Emirates	MetaDubai, Ripple
France	Stage11
Germany	RIMOWA
Italy	Gucci

3 Metaverse and Tourism

The term "metaverse" in the context of the tourism industry refers to a 3D virtual environment that facilitates social interactions between travellers, travel agencies, and a range of stakeholders. This offers opportunities to establish novel means of delivering services by connecting the virtual and physical realms of the metaverse and the real world (Koo et al. 2022). A lasting network of communal virtual spaces, known as the metaverse, offers individuals a full sense of existence by means of avatars and harmonises mental and physical processes during concurrent interaction with other entities and objects. It is suggested that metaverse excursions are a new development that the travel and leisure sector must fully understand (Tsai 2022). It allows individuals to participate in such an experience as an avatar and share their journey adventures in a virtual setting via social media (Um et al. 2022). The tourism sector's metaverse combines real-life with mixed reality to unify all requirements and stakeholders in a three-dimensional virtual realm. It alters tangible spaces into mixed reality spaces, creating a parallel virtual universe on the internet (Buhalis and Karatay 2022).

In the travel industry, technological advancements have significantly impacted customer behaviour (Wei 2022; Atiyah, 2023). Like other sectors, the travel and tourism industry has been influenced by technology over time. For example, virtual reality - which is expected to become increasingly commonplace in the future - can provide a range of experiences through computer - generated visual and graphic applications (Çolakoğlu et al. 2023; AL-Fatlawey et al. 2021). In this context, the tourism industry has implemented concepts such as "Metatourism" and "Metahotels" with the use of metaverse technology. Meta tourism, as an umbrella term, enables people to experience overseas travel without actually leaving their homes (Arasa 2022). Conversely, Metahotels refer to digital hotels that provide personalized experiences to individuals through avatars at their homes (Demir 2022). As travel agencies facilitate the sale of products or services via virtual platforms, the concept of metagencies or metaoperators can be easily applied in this context. This notion is reinforced by Babur's (2022) assertion that the metaverse will transform the tourism and travel industry in the upcoming years. From this perspective, it is crucial to understand how people respond to the emergence of new technology (Ketikidis et al. 2012).

According to Go and Kang (2023), tourists' experiences with a tourism destination or product can be enhanced through interaction with virtual reality environments. Considering the potential versatility of the metaverse (for instance, booking, decision-making, transformative consumption, co-creation), it is projected to become a powerful technological foundation for the travel industry (Wei 2023; Khaw et al., 2022). According to a study conducted by Monaco and Sacchi (2023), many digital solutions were commonly adopted in the tourism sector prior to the pandemic, but the metaverse might introduce new dimensions that can enhance tourism opportunities beyond physical locations. Lockdowns have prompted individuals to seek out unconventional and alternative transportation methods, inspiring players within the travel and tourism industry to offer distinctive experiences. These cater to tourists' desire for activities promoting safety and sustainability, enabling them to travel virtually across the globe without physically journeying there. Besides, tourists have the option to leave their identities in the physical realm, thus enabling the construction of a multi-identity experience in the virtual world (Koo et al. 2022).

When metaverse studies are examined, it is observed that a conceptual framework about the metaverse is attempted to be created (Murti et al. 2023; Çolakoğlu et al. 2023; Rather 2023; Wei 2023; Chen et al. 2023; Buhalis et al. 2023; Gursoy et al. 2022; Volchek and Brysch 2022), the relationship between the metaverse and tourism is evaluated, particularly in terms of global crises and sustainability (Monaco and Sacchi 2023, Go and Kang 2023; Wei 2022; Zaman et al. 2022; Suanpang et al. 2022), and possible changes in the industry's future, consumer behaviour, decision process, and consumer habits are emphasized rather than tourism suppliers (Murti et al. 2023; Huang et al. 2023; Koo et al. 2022; Choi and Kim 2017; Navarro 2013; Gomes and Araujo 2012). Indeed, Chen et al. (2023) state that most research on the topic has concentrated on the theoretical framework of metaverse tourism and its impact on tourist behaviour, tourism operations, and tourism management, despite its essential function in tourist experiences and tourism management. Therefore, exploring the perspectives of industry professionals is considered necessary for understanding the industry implications of the metaverse. Although the system adapts to meet customer needs, it is important to consider the opinions of employees involved in the system. Travel agencies, which act as a bridge between producers and customers, are the foundational components of this industry and create products within this structure. As emerging tourism technologies are implemented, these agencies become key players in the development process. Based on the findings, this study offers guidance for travel agency employees to seize the opportunity to develop the metaverse and fully exploit its potential.

4 Method

Research on the consequences and contributing features of metaverse tourism is limited due to the unprecedented impact the metaverse has on various dimensions, including tourism (Chen et al. 2023). Qualitative research within this context can reveal new perspectives and integrations, aiding in surpassing the conceptions and frameworks developed in the early stages of the research (Miles and Huberman 1994). For this reason, qualitative research methods were utilised to investigate a novel field as a research methodology and to interpret the research from a holistic viewpoint based on the perspectives of the participants. This approach facilitated a comprehensive exploration of subjective situations rather than seeking a singular truth and aimed to capture multiple perspectives in a particular context (Çolakoğlu et al. 2023).

The objective of this research is to investigate the perceptions of Metaverse among travel agency employees and its potential impact on the enterprises. While the tourism industry primarily considers the impact of the metaverse on services and consumers, the views of tourism stakeholders are frequently overlooked. The relevant literature has been thoroughly reviewed, incorporating sample studies about Metaverse and tourism. To develop the research questions, previous studies from the literature were referenced (Álvarez et al. 2007; Buhalis and Law 2008; Dionisio et al. 2013; Tayfun et al. 2022). The interview questionnaire, consisting of two sections and eleven questions, was developed with expert assistance and administered to the participants. The initial section contained questions to measure travel agency employees' perceptions of the metaverse, while the following section examined the potential impact on the travel industry. Focus groups were surveyed using the interview method, a qualitative data collection method. Descriptive and content analysis were employed to analyse the data obtained.

The interviews were recorded without any personal questioning or time restrictions. The average interview duration was 45 min. Before the actual interview, the participants were given a brief on the topic. The study's sample consisted of nine managers and staff from travel agencies operating in Istanbul and was selected using the convenience sampling method. The participants work in various business departments: 1 company owner, 4 managers, 1 reservation specialist, 1 customer relations representative, 1 sales marketing coordinator, and 1 product development manager.

4.1 Research Questions

Part One: Metaverse

- What major technological advances have occurred in the travel and tourism industry over the past ten years? What impact have technological advances had on tourism?
- Are you familiar with metaverse? If so, can you describe it as you would see it?
- Would you think about investing personally in Metaverse? Have you made a Bitcoin investment?
- Do you follow companies that finance Metaverse? Are investments in this field wise in your opinion?
- What do you think about the connection between the metaverse and tourism?

Part Two: Potential Effects of Metaverse

- Do you consider the idea of the metaverse to be a tool for marketing? How does it fit into marketing?
- What kinds of improvements do you anticipate occurring in the metaverse in terms of marketing initiatives for your company?
- What alterations do you expect the Metaverse to bring about regarding the tourist experience?
- What are the benefits and drawbacks of the application of the metaverse idea to the travel industry?
- What advantages and disadvantages do tourists stand to reap from the development of the metaverse concept?
- What do you believe will happen to the idea of the metaverse? As a company, are you prepared for these changes and innovations?

4.2 Findings

Findings According to the Demographic Characteristics of the Participants.

The demographic and occupational details of the nine participants in the study are shown in Table 2.

Table 2. Demographic and Occupational Characteristics of the Participants.

Demographic Characteristic		f	Occupational Characteristics	f
Gender	Female	4	**Work Experience**	
	Male	5	6–10 years	2
Age	26–35	2	11–15 years	1
	36–45	5	16–20 years	2
	46–55	2	20 years and above	4
			Graduated University Department	
			Travel Management	5
			Tourism and Hospitality Management	4

Upon examination of the demographic data presented in Table 2, it becomes evident that the sample consists of 4 male and 5 female participants, with the majority falling within the age range of 36 to 45. It can be noticed from the participants' educational backgrounds that they were all travel and tourism department graduates. The participants' experience period is at least 6 years, according to the findings regarding the participants' professional characteristics.

4.3 Findings of the Part One: Metaverse

In the initial phase of the study, inquiries were made to ascertain the sentiments of the participants regarding the concept of the metaverse. "What major technological advances have taken place in the travel and tourism industry over the past ten years? What impact have technological advances had on tourism?" responses to the question below:

P3: *'Things were quite different when I started my first job. For instance, we were faxing hotel reservations. Then, of course, it was sent to the mail. As reservation systems evolved, the data we manually entered became available through these systems. Every piece of equipment at my current job is linked to the system. My reservation goes directly to the hotel when I do so. Furthermore, the guest and I both receive confirmations once the reservation is complete. The system can simply provide transactions for payments and invoices. Also, tourists have online access to all of this data. In other words, whereas initially only 50 reservations could be made, it is now quite simple to make 150 or more reservations. The workload is reduced as a result, and transactions move along more quickly. Additionally, visitors can now use our system to make their own reservations."*

P7: ''*New products did not, in reality, start to exist with technology. There were innovations in promotion and marketing. Vendors can now directly contact consumers, as opposed to previously having to find consumer vendors. For instance, communication and reservation systems used to be conducted manually over the phone and fax. The mutual error rates were higher, as expected. Consumer mistrust was a result of these errors. Thanks to new technology, digitalisation has accelerated processes, reduced margins for error, and increased satisfaction.*

P9: ''*First of all, I don't think we can effectively integrate technological advancements into travel or tourism as a country. We observe that in other countries like the Far East, people use technology better. I do, however, notice the attempts made by agencies, hoteliers, or other businesses to innovate and use technology more effectively, particularly in the wake of the pandemic process.*

P1: ''*Middle Eastern in origin, the company I work for still lags behind in some technological areas. While other agencies use data to communicate, we continue to use WhatsApp. In other words, no matter how much technology advances, it won't make any sense if people and the market don't adapt to it.*

Second, "Are you familiar with metaverse? If so, can you describe it as you would see it?" was the query. According to the findings, each participant had heard of the Metaverse concept before but did not have extensive knowledge. The participants mainly focused on the notions of "virtual world" (f = 16), "imaginary world" (f = 4), and "virtual state of the existing world" (f = 2). Figure 1 depicts the participants' Metaverse definitions as a word cloud and concepts.

In order to measure the perceptions of the participants regarding metaverse investments, ''Would you think about investing personally in Metaverse? Have you made a Bitcoin investment? Do you follow companies that finance Metaverse? Are investments in this field wise in your opinion?" questions were posed. Table 3 represents the participants' virtual investment status and investment plans.

Upon examination of the participants' reactions to the metaverse investment, it becomes evident that a majority of them exhibit enthusiasm towards the situation, albeit with a subset lacking sufficient knowledge on the subject matter. Currently, the number of people investing in Bitcoin, one of the virtual currency types, is 4.

P3: *"I'm not very interested due to my age, but if someone from the younger generation informs and guides me, why not?".*

Fig. 1. Metaverse Definitions of Participants

Table 3. Participants' Perceptions of Metaverse Investment.

Metaverse Investment Plan	F
Yes	5
No	4
Bitcoin Investment Status	F
Yes	4
No	5

P5: *"I am interested in Bitcoin and other virtual currencies. I had a pre-pandemic experience, so I only had a passing thought about it because I didn't have much knowledge. As you may be aware, when certain agencies were closed due to the pandemic, I began to fall for the job a little more on the Internet. As a result, it became a source of income throughout the pandemic, and I used it to cover all of my agency's expenses. I continue to believe in the validity of virtual currency. In the following years, I will most likely continue to invest. I've heard about Metaverse but have never invested. This is solely due to a lack of time; if given the opportunity, I would surely invest."*

Six participants state that they keep up with the innovations in the metaverse field and think this is wise.

P9: *"I'm on board; it's sound advice. We put in the practice to shoot Titanic Mardan Palace."*

P1: *"Facebook has made a huge investment. Even the company's name was changed. Nike and Adidas, for example, are investing in various apps, and I keep an eye on them. In fact, I know that metaverse funds have begun to be established in some co-operations."*

P4: *"We notice that people purchased something in the metaverse, that they had a wedding, and we follow it in the news. However, I do not see companies doing much about it. Only in the field of marketing and advertising in global corporations do I believe there are applications to this extent."*

The first part's final question, "What do you think about the connection between the metaverse and tourism?" has been determined. The responses to the question were coded using the themes "experience" and "marketing." Accordingly, participants in the Metaverse and tourism relationship generally express this relationship through marketing (f = 7), with only two persons explaining it through experience.

P7: *"I look at it from a marketing standpoint. Hotels started making 360-degree videos to sell themselves over a decade ago. They started with Instagram, and now they're employing drones. They will begin implementing Metaverse in the next phase. This will be the definitive meeting of the metaverse and tourism."*

P1: *"It's something I'm actually interested in. I've been thinking about the manner in which the metaverse can be used in tourism for a long time. I even saw certain developments, such as the opening of virtual stores. In my perspective, it could be used for more activities. It can be quite handy for congress tourism, for example. Chat rooms with speakers can be found and participants can attend these events with an online ticket."*

4.4 Findings of the Part Two: Potential Effects of Metaverse

In the second part of the research, participants were initially asked questions regarding the metaverse and marketing. The questions are ''Do you consider the idea of the metaverse to be a tool for marketing? How does it fit into marketing? What kinds of improvements do you anticipate occurring in the metaverse in terms of marketing initiatives for your company?" was directed to the participants in accordance with the march of interview. According to the responses, all participants consider metaverse as a marketing tool. However, some participants claim this is a process and only progresses slowly. According to one participant who has used metaverse for marketing objectives, this technology is an innovation that will take time due to budgetary constraints and infrastructure issues.

P6: *"It can be used in marketing, of course. Information can be supported by digital devices. If a seller in the agency shows the places for children from the computer and delivers an experience to the child while wearing VR glasses, it will boost the conversion of this information into sales."*

P8: *"It will undoubtedly be used in the future. We are now solely promoting it in Dubai. We let people experience it firsthand. Of course, the response was overwhelmingly positive. I'm also among those who rated it positively because I used it for the first time as well. It was very realistic indeed. People are passing by you at that moment, you feel. I'm quite impressed, yet it's just too costly right now. Shooting those videos, for example, requires time and is fairly technical. But, as it becomes more popular, I believe the costs will fall and more people will use it."*

Another question posed to participants was, ''what alterations do you expect the Metaverse will bring about in terms of the tourist experience?". Most of the responses (f = 7) indicate that Metaverse technology will impact marketing rather than experience. Two participants believe that this technology will be used in tourist experiences in the future, particularly for individuals concerned about time and financial loss. They also point out that it may allow individuals with disabilities to engage in tourism activities.

P3: *"I look at it from a marketing perspective as opposed to experience. Because people should taste and smell the food they see, I believe it is difficult for them to feel it. Of course, we have no idea how it will evolve in the future. So I don't think it will have an immediate impact on the experiences anytime soon."*

P5: *''There is a true experience when we talk about tourism. Interaction, living, experiencing reality, I don't know, touching, tasting should be all. When visiting Spain, you must smell the flavour of the cuisine you eat. If you do this virtually, it seems like you will not have the same experience or be able to taste the same. Tourism fulfils people in terms of travelling and seeing, but it is not sufficient; you must also live with the conditions of that destination. Tourism is more than just places to visit. In my opinion, getting on a bus from that country and eating the food from that country should be experienced. You can achieve that to a limited extent with technology."*

P1: *"I believe we can visit plenty of places from our house. Virtual reality glasses can help with this. We may not be able to experience, but young people will be able to.".*

In the next stage, the participants were then asked to assess the metaverse technology in terms of tourists and travel agencies. Table 4 summarizes the positive and negative aspects of metaverse technology as mentioned by participants.

Table 4. Advantages and Disadvantages of Metaverse Technology

Advantages of Metaverse for Tourists	**Advantages of Metaverse for Travel Agencies**
Opportunity to access the service without consuming	Opportunity to reach more people
Reduction of pecuniary loss and intangible damages	Realistic and efficient marketing
Elimination of time and cost constraints	Increasing seller motivation
Encouraging tourists	Reduced complaints as a result of pre-experience
A viable experience option for the disabled tourists	
Disadvantages of Metaverse for Tourists	**Disadvantages of Metaverse for Travel Agencies**
Pre-experience and experience do not overlap	Pre-experience and experience do not overlap
	Direct access of tourists to service providers

Finally, the question, "What do you believe will happen to the idea of the metaverse? As a company, are you prepared for these changes and innovations?" was posed. All participants agree that the metaverse is a developing technology that will take time to mature. Furthermore, some participants believe that this technology, like the internet, will become widely used in the future. According to the responses gathered, no investment in the issue has yet been made in any travel agencies. However, based on the interviews, travel agencies are open to fresh ideas and are willing to invest in the right moment and situation.

P4: '*The metaverse evolves with its users, and it has the potential to become as widespread as the internet. I don't have an adverse assessment, especially because we adapt swiftly to emerging technologies as a country. When I was in college, there were push-button phones. Then we shifted to touch phones, and we were able to react quickly to these alterations."*

P2: *"Metaverse is a futuristic technology. I believe it will be used more frequently and widely. I am aware that some multinational hotel chains (such as Hilton) are interested. To be honest, I believe the first ones, those without a cost issue; will receive far more positive feedback. But, of course, it will be a lengthy process."*

P9: *"I believe it is a world with a future. Because, like it or not, we need to keep up with technology. We are in that circle, and do not have the luxury of getting out of it. We are not financially prepared as a company. I would definitely take the initiative if it was a grant or a joint project. It is on our agenda, and we are following it.*

P7: *"I believe it will fail numerous times. It will be centred on constant progress, and the intended outcome will be realized eventually. I'm optimistic that it will expand significantly. At the moment, there is no substantial investment in our company, but I think that in the future, other businesses will follow in the footsteps of large corporations such as Google."*

5 Discussion

The metaverse concept, whose effects began to be observed more clearly in 2022, is recognised as an invention that could impact several industries (Dwivedi et al., 2023; AL-Abrrow et al., 2023). According to the World Economic Forum, tourism is one of the significant industries that may benefit from the metaverse (Buhalis et al., 2023). In the marketing domain, the ideas of metaverse and tourism are often combined (eMarketer, 2021; Hollensen et al., 2022; Buhalis et al., 2023). Accordingly, the second stage of the research included investigations into marketing.

In the initial phase of the study, the goal was to assess the participants' perceptions of the metaverse. The results indicate that although the participants had heard of the term before, they had not explored it due to a lack of comprehensive information. Duran et al. (2022) elucidate this situation using the "Gray Rhino" metaphor in their investigation. As per the study, the "metaverse" concept is akin to an "unidentified gray rhino". The unidentified grey rhino appears to be hesitant and indecisive in taking action, despite being aware of the situation.

The aim of the second section of this research is to determine the influence and potential benefits of the concept of a metaverse in tourism marketing. Based on the responses of the participants, the results demonstrate that the advantages of the metaverse exceed its disadvantages. This outcome is in line with previous research conducted on the topic. The advantages of the metaverse on the tourism sector, as per the findings of Kaya et al.'s (2023) research, comprise of events that are virtual or a blend of virtual and physical ones, virtual voyages to far-off and novel destinations, business trips and meetings that are virtually attended, virtual tours to hotels, memorable tourist experiences that are immersive, sustainability, augmented reality-based experiences for shopping and menus, and employee experiences. Employee experience, cyber security and law, accessibility, and sustainability were all cited as drawbacks in a single study. Conversely, another study focused exclusively on the potential benefits and prospects of the metaverse. Potential advantages of the metaverse for the travel and tourism industry are stated as being sales-oriented, such as promoting travel-related buying patterns, enhancing the booking experience, and boosting booking volume (Revfine 2022).

However, only a portion of the metaverse tourism ecosystem has yet to be implemented thus far. There are many restrictions, such as improper technology (such HMDs), inadequate integration, and limited affectivity (such as taste, touch, and smell). To create immersive experiences, to manipulate travel behaviour, sensory-rich environments are required (Koo et al. 2022). The metaverse, on the other hand, is anticipated to encourage travel behaviour more. Individuals who engage in virtual travel experiences and tourism-related activities may also be more likely to utilize physical travel-related products and visit the destination. After virtual visits, users can share content on platforms and continue interacting with other individuals they meet (Buhalis et al. 2023). Notably, alternative digital or virtual tourism platforms can be provided, especially to tech-savvy tourists or those concerned about travel (Zaman et al. 2022). Participants have expressed concerns about the transformation of both experiences and services within the tourism industry. While it is uncertain how the future will take shape, the experiential nature of tourism services is seen as a potential issue.

The metaverse is widely regarded as the next major development. Developing countries face several obstacles to reap full benefits from the metaverse (Kshetri 2022). If tourism professionals are not familiar with technology and digital platforms, it may be challenging to attain success (Yemenici 2022). Furthermore, the uncertain development direction, technological barriers, and resistance to change pose significant obstacles for travel agencies to adopt the metaverse in the tourism industry. Marketing through the Metaverse is experimental and presents multiple challenges for suppliers, including technological and infrastructure considerations, socio-cultural issues, and development strategies (Chen et al. 2023). Despite the challenges faced, the participants hold the belief that the metaverse represents the future technology. The lag behind evolving technologies poses an issue, particularly impeded by inadequate infrastructure and financing. Although the participants have positive viewpoints regarding the metaverse, they hold reservations about its suitability to the tourism industry. This extends beyond the country; both the market and business structure may influence the perception of the metaverse.

6 Conclusion and Recommendation

The study's findings reveal that employees in the industry assess the Metaverse idea and its potential repercussions for the tourism industry from various approaches. The subject still needs to be solved, highlighted by the absence of agreement on the definition and the appropriate field to be used. Nevertheless, enterprises that embrace and adjust to advancements in a perpetually evolving world consistently possess a competitive edge over their competitors. Investments in the metaverse and related technologies are growing in this setting. One of the tangible steps achieved for the metaverse technology may be seen in the glasses that Apple unveiled as the three-dimensional Vision Pro. One of the fundamental instruments for the metaverse, which fuses the real and virtual worlds, is anticipated to be these glasses, which keep both of them together. This technology seems far off and is gradually beginning to materialize in the real world. Based on this, stakeholders in the industry must decide where they fit into this technology.

A completely new commerce industry might be generated due to the metaverse and the possibilities of virtual travel. Therefore, in order to keep up with new technologies, conventional online travel agencies might require modifying their business strategies. Expanded possibilities for flexible travel, customized consumer services, and entertainment might all be a part of the metaverse's virtual evolution. The metaverse can offer seamless interaction and immersive experiences between users in real and simulated environments by applying augmented and virtual reality technology. Tourists can bypass the retailers and contact the supplier directly for real-time price changes, availability and promotions. At this point, the focus should be on leveraging sensory signals to affect visitors' purchasing decisions and product preferences in a virtual environment. Emotional and sensory signals are required in virtual formations, just as physical cues are in actual tourism activities. The conventional tourism marketing mix could shift entirely as a result of this. Encouraging the participation of disadvantaged individuals, such as the physically challenged, in tourism-related activities may establish new market segments.

The metaverse will change tourism in the future, and travel agencies must be prepared for both the good and bad repercussions. This continuum may bring both opportunities and difficulties for them. In this context, the metaverse may not only improve tourist experiences, but the merging of reality and virtuality may also alter definition of the paradigms, ethics, values, norms, standards, experiences, and business practices ways of doing.

Research on this topic is groundbreaking, as studying the metaverse in tourism is a novel area. To encourage increased investment in the metaverse, it is vital that industry professionals possess a thorough comprehension of the topic. Thus, it may be beneficial to create various tools to evaluate employees' attitudes within the industry. While qualitative research offers beneficial and all-inclusive perspectives, it is limited by certain constraints that restrict the applicability of its findings. Additionally, this study's limited scope only encompasses one particular region, meaning the opinions of tourism professionals from differing regions and countries on the metaverse may differ. In addition, it is important to consider the varying levels of technological advancement between developing nations and technologically developed countries that produce, implement and market technology, as this may affect their opinions on the metaverse. The study is restricted in its scope to only cover travel agencies; hence other tourism stakeholders may have different degrees of technological integration. To remedy this, further investigations should be conducted using a diverse range of sample groups.

6.1 Theoretical Implications

Research into the metaverse is generally considered to be an assessment of theoretical concepts (Murti et al. 2023; Çolakoğlu et al. 2023; Rather 2023; Wei 2023; Chen et al. 2023; Buhalis et al. 2023; Gursoy et al. 2022; Volchek and Brysch, 2022). The initial theoretical contribution of this investigation is to broaden research on metaverse tourism beyond the conceptual stage to practical exploration. Buhalis et al. (2023) argue that comprehending the metaverse and creating practicable and beneficial platforms, procedures, and services can enhance the co-creation of value for all interested parties. The second contribution of this study is to investigate metaverse perspectives, based on opinions from travel agency employees who are key stakeholders in tourism. It is important to balance their views to ensure successful development of Metaverse tourism. The study concludes by providing a theoretical foundation for future research.

6.2 Practical Implications

Stakeholders' views can offer insight into the utilization, expansion, possibilities and risks of the metaverse in the tourism industry. In this context, the need for various regulations is apparent. The government, as the primary stakeholder in metaverse tourism, has the potential to engage all stakeholders in the development of this sector. In this context, soliciting opinions from multiple parties and understanding their needs is crucial to promote meta-tourism in an effective and efficient way, while also taking necessary precautions (Nyanjom et al. 2018).

A stable economic and social environment is essential for the existence of tourism-related activities. However, crises such as epidemics undermine the confidence of consumers who make travel decisions and discretionary purchases (Page and Yeoman, 2006). The recent Covid-19 crisis serves as a significant example of this. Within this context, individuals who are disadvantaged and unable to travel for various reasons, or who are fearful of doing so, can still engage in tourism activities without the need for physical mobility. In this context, the successful implementation of the metaverse will make a substantial contribution to the tourism industry. New tourism offerings and adventures can be fashioned with the backing of technology (Şahin and Güven 2022). Thus, both the tourist experience can be enhanced and the product range of the enterprises can be broadened.

References

Abbas, S., et al.: Antecedents of trustworthiness of social commerce platforms: a case of rural communities using multi group SEM & MCDM methods. In: Electronic Commerce Research and Applications, 101322 (2023)

AL-Abrrow, H., et al.: Understanding employees' responses to the COVID-19 pandemic: the attractiveness of healthcare jobs. Global Bus. Organ. Excell. **40**(2), 19–33 (2021)

AL-Fatlawey, M.H., Brias, A.K., Atiyah, A.G.: The role of Strategic Behavior in achievement the Organizational Excellence "Analytical research of the manager's views of Ur State Company at Thi-Qar Governorate". J. Adm. Econ. **10**(37) (2021)

Al-Hchaimi, A.A.J., Sulaiman, N.B., Mustafa, M.A.B., Mohtar, M.N.B., Hassan, S.L.B.M., Muhsen, Y.R.: A comprehensive evaluation approach for efficient countermeasure techniques against timing side-channel attack on MPSoC-based IoT using multi-criteria decision-making methods. Egyptian Inf. J. **24**(2), 351–364 (2023)

Alnoor, A.M., Al-Abrrow, H., Abdullah, H., Abbas, S.: The impact of self-efficacy on employees' ability to accept new technology in an Iraqi university. Glob. Bus. Organ. Excell. **39**(2), 41–50 (2020)

Alnoor, A., et al.: How positive and negative electronic word of mouth (eWOM) affects customers' intention to use social commerce? A dual-stage multi group-SEM and ANN analysis. International Journal of Human–Computer Interaction, pp. 1–30 (2022)
Álvarez, L.S., Martín, A.M., Casielles, R.V.: Relationship marketing and information and communication technologies: analysis of retail travel agencies. J. Travel Res. **45**(4), 453–463 (2007)
Ane, B.K., Roller, D., Lolugu, J.: Ubiquitous virtual reality: the state-of-the-art. Int. J. Comput. Sci. Mob. Comput. **8**(7), 16–26 (2019)
Arasa, D. 6 February 2022. https://usa.inquirer.net/91663/metaverse-tourism
Atiyah, A.G.: Impact of knowledge workers characteristics in promoting organizational creativity: an applied study in a sample of smart organizations. PalArch's J. Archaeol. Egypt/Egyptol. **17**(6), 16626–16637 (2020)
Atiyah, A.G.: Power distance and strategic decision implementation: exploring the moderative influence of organizational context
Atshan, N.A., Al-Abrrow, H., Abdullah, H.O., Khaw, K.W., Alnoor, A., Abbas, S.: The effect of perceived organizational politics on responses to job dissatisfaction: the moderating roles of self-efficacy and political skill. Glob. Bus. Organ. Excell. **41**(2), 43–54 (2022)
Aymen, R.A., Alhamzah, A., Bilal, E.: A multi-level study of influence knowledge management small and medium enterprises. Polish J. Manag. Stud. **19**(1), 21–31 (2019)
Babur, Y.: Metaverse ağında turizm sektörünün rolü. Turizm Ekonomi ve İşletme Araştırmaları Dergisi **4**(1), 91–101 (2022)
Bozanic, D., Tešić, D., Puška, A., Štilić, A., Muhsen, Y.R.: Ranking challenges, risks and threats using Fuzzy Inference System. Dec. Making Appl. Manage. Eng. **6**(2), 933–947 (2023)
Buhalis, D., Karatay, N.: Mixed reality (MR) for generation Z in cultural heritage tourism towards metaverse. In: Stienmetz, J.L., Ferrer-Rosell, B., Massimo, D. (eds.) ENTER 2022, pp. 16–27. Springer, Cham (2022). https://doi.org/10.1007/978-3-030-94751-4_2
Buhalis, D., Law, R.: Progress in information technology and tourism management: 20 years on and 10 years after the Internet—the state of eTourism research. Tour. Manage. **29**(4), 609–623 (2008)
Buhalis, D., O'Connor, P.: Information communication technology revolutionizing tourism. Tour. Recreat. Res. **30**(3), 7–16 (2005)
Buhalis, D., Leung, D., Lin, M.: Metaverse as a disruptive technology revolutionising tourism management and marketing. Tour. Manage. **97**, 1–11 (2023)
Chen, S., Chan, I.C., Xu, S., Law, R., Zhang, M.: Metaverse in tourism: drivers and hindrances from stakeholders' perspective. J. Travel Tour. Mark. **2**, 169–184 (2023)
Chew, X., Khaw, K.W., Alnoor, A., Ferasso, M., Al Halbusi, H., Muhsen, Y.R.: Circular economy of medical waste: novel intelligent medical waste management framework based on extension linear Diophantine fuzzy FDOSM and neural network approach. Environ. Sci. Pollut. Res. 1–27 (2023)
Choi, H.-S., Kim, S.-H.: A content service deployment plan for metaverse museum exhibitions—Centering on the combination of beacons and HMDs. Int. J. Inf. Manage. **37**, 1519–1527 (2017)
Coelho, C., Tichon, J., Hine, T., Wallis, G., Riva, G.: Media presense and inner presence: the sense of presence in virtual reality technologies. In: Riva, G., Anguera, M.T., Wiederhold, B.K., Mantovani, F. (eds.) From Communication to Presense: Cognition, Emotions, Culture Towards the Ultimate Communicative Experience, pp. 26–45. IOS, Amsterdam (2006)
Çolakoğlu, Ü., Anış, E., Esen, Ö., Tuncay, C.S.: The evaluation of tourists' virtual reality experiences in the transition process to Metaverse. J. Hospitality Tour. Insights **0**(0), 1–26 (2023)
Cranmer, E.E., Dieck, M.C., Fountoulaki, P.: Exploring the value of augmented reality for tourism. Tourism Manage. Perspect. **35**(4), 1–9 (2020)

Dahan, N.A., Al-Razgan, M., Al-Laith, A., Alsouf, M.A., Al-Asaly, M.S., Alfakih, T.: Metaverse framework: a case study on e-learning. Electronics **11**(10), 1–13 (2022)

Davis, A., Murphy, J.D., Owens, D., Khazanch, D., Zigurs, I.: Avatars, people, and virtual worlds: Foundations for research in metaverses. J. Assoc. Inf. Syst. **10**(2), 91–117 (2009)

Demir, Ç.: Metaverse Teknolojisinin Otel Sektörünün Geleceğine Etkileri Üzerine Bir İnceleme. J. Tourism Gastron. Stud. **10**(1), 542–555 (2022)

Dionisio, J.D., Burns, W.G., Gilbert, R.: 3D virtual worlds and the metaverse: current status and future possibilities. ACM Comput. Surv. **45**(3), 34–73 (2013)

Dolata, M., Schwabe, G.: What is the Metaverse and who seeks to define it? Mapping the site of social construction. J. Inf. Technol. **0**(0), 1–28 (2023)

Donthu, N.: Effects of COVID-19 on business and research. J. Bus. Res. **117**, 284–289 (2020)

Duran, G., Kanıgür, S., Hassan, A.: Gri gergedan metaforu bağlamında metaverse'ün turizm sektörü açısından incelenmesi. J. New Tourism Trends **3**(2), 160–176 (2022)

Dwivedi, Y., et al.: Metaverse beyond the hype: Multidisciplinary perspectives on emerging challenges, opportunities, and agenda for research, practice and policy. Int. J. Inf. Manage. **66**(C), 1–55 (2022)

Emarketer: Defining the metaverse, when advertisers should jump in, and rethinking time (2021). https://www.Emarketer.Com/Content/Podcast-Defining-metaverse-advertisers-should-jump-in-rethinking-time?Ecid=NL1009

Ercan, F.: Metaverse teknolojisinin gelecekte turizm sektörüne olası etkilerini belirlemeye yönelik bir araştırma. Anadolu Üniversitesi Sosyal Bilimler Dergisi **22**(4), 1063–1092 (2022)

Gatea, A.A., Marina, V.: Higher education funding in Iraq in terms of the experience of particular developed countries. Int. J. Adv. Stud. **6**(1), 8–17 (2016)

Go, H., Kang, M.: Metaverse tourism for sustainable tourism development: Tourism Agenda 2030. Tourism Rev. **78**(2), 381–394 (2023)

Gomes, D.A., Araujo, M.C.: Tourism online: a metaverse offers study. Estudios y Perspectivas en Turismo **21**(4), 876–903 (2012)

Guduraš, D.: Economic crisis and tourism: case of the Greek tourism sector. Ekon. Misao Praksa Dbk. God **23**(2), 613–632 (2014)

Gursoy, D., Malodia, S., Dhir, A.: The metaverse in the hospitality and tourism industry: an overview of current trends and future research directions. J. Hospit. Mark. Manage. Rev. **31**(5), 527–534 (2022)

Han, D.-I., Jung, T., Gibson, A.: Dublin AR: implementing augmented reality in Tourism. In: Xiang, Z., Tussyadiah, I. (eds.) Information and Communication Technologies in Tourism 2014, pp. 511–523. Springer, Cham (2013). https://doi.org/10.1007/978-3-319-03973-2_37

He, Z., Wu, L., Li, X.: When art meets tech: the role of augmented reality in enhancing museum experiences and purchase intentions. Tour. Manage. **68**, 127–139 (2018)

Helsel, S.: Virtual reality and education. Educ. Technol. **32**(5), 38–42 (1992)

Hoffman, H.G.: Virtual reality therapy. Sci. Am. **291**(2), 58–65 (2004)

Hollensen, S., Kotler, P., Opresnik, M.: Metaverse - the new marketing universe. J. Bus. Strateg. **44**(3), 119–125 (2022)

Huang, X.-T., Wang, J., Wang, Z., Wang, L., Cheng, C.: Experimental study on the influence of virtual tourism spatial situation on the tourists' temperature comfort in the context of metaverse. Environ. Psychol. **13**, 1–16 (2023)

Jung, S.H., Jeon, I.-O.: A study on the components of the Metaverse ecosystem. J. Digit. Conv. **20**(2), 163–174 (2022)

Kaya, B., Bayar, S.B., Meydan Uygur, S.: Is metaverse a comfortable parallel universe for the tourism industry or a nightmare full of fear?. Univ. South Florida (USF) M3 Publishing, **2**, 1–9 (2022)

Ketikidis, P., Dimitrovski, T., Lazuras, L., Bath, P.A.: Acceptance of health information technology in health professionals: an application of the revised technology acceptance model. Health Informatics J. **18**(2), 124–134 (2012)

Khaw, K.W., Alnoor, A., Al-Abrrow, H., Tiberius, V., Ganesan, Y., Atshan, N.A.: Reactions towards organizational change: a systematic literature review. Curr. Psychol. 1–24 (2022)

Kim, M.J., Hall, C.M.: A hedonic motivation model in virtual reality tourism: comparing visitors and non-visitors. Int. J. Inf. Manage. **46**(4), 236–249 (2019)

Koo, C., Kwon, J., Chung, N., Kim, J.: Metaverse tourism: conceptual framework and research propositions. Curr. Issues Tour. **0**(0), 1–8 (2022)

Kshetri, N.: Metaverse and developing economies. It Econ. **24**(4), 66–69 (2022)

Kwok, A.O., Koh, S.G.: COVID-19 and extended reality (XR). Curr. Issue Tour. **24**(14), 1935–1940 (2021)

Kye, B., Han, N., Kim, E., Park, Y., Jo, S.: Educational applications of metaverse: possibilities and limitations. J. Educ. Eval. Health Prof. **18**(32), 1–13 (2021)

Lee, L.-H., et al.: All one needs to know about metaverse: a complete survey on technological singularity, virtual ecosystem, and research agenda. J. X Class Files **14**(8), 1–66 (2021)

Lu, J., Xiao, X., Xu, Z., Wang, C., Zhang, M., Zhou, Y.: The potential of virtual tourism in the recovery of tourism industry during the COVID-19 pandemic. Curr. Issue Tour. **25**(3), 441–457 (2021)

Manhal, M., Al-khalidi, A., Hamad, Z.: Strategic network: managerial myopia point of view. Manage. Sci. Lett. **13**(3), 211–218 (2023)

McKinsey & Company: Value Creation in the Metaverse. McKinsey & Company, Chicago (2022)

Miles, M.B., Huberman, A.M.: Qualitative Data Anlaysis. Sage, California (1994)

Mohanty, P., Hassan, A., Ekis, E.: Augmented reality for relaunching tourism post COVID-19: socially distant, virtually connected. Worldwide Hospit. Tour. Themes **12**(6), 753–760 (2020)

Monaco, S., Sacchi, G.: Travelling the metaverse: potential benefits and main challenges for tourism sectors and research applications. Sustainability **15**(4), 1–10 (2023)

Morimoto, T., et al.: XR (Extended reality: Virtual reality, augmented reality, mixed Reality) technology in spine medicine: status quo and quo vadis. J. Clin. Med. **11**(470), 1–23 (2022)

Muhsen, Y.R., Husin, N.A., Zolkepli, M.B., Manshor, N., Al-Hchaimi, A.A.J.: Evaluation of the routing algorithms for NoC-Based MPSoC: a fuzzy multi-criteria decision-making approach. IEEE Access (2023)

Mystakidis, S.: Metaverse. Encyclopedia **2**, 486–497 (2022)

Narin, N.G.: A content analysis of the metaverse articles. J. Metaverse **1**(1), 17–24 (2021)

Navarro, S.: Virtual environment and tourism Symbolic interaction. Revista Iberoamericana de Turismo **3**(2), 17–24 (2013)

Ning, H., et al.: A survey on metaverse: the State-of-the-art technologies, applications, and challenges. Cornell University (2023). https://arxiv.org/ftp/arxiv/papers/2111/2111.09673.pdf

Nyanjom, J., Boxall, K., Slaven, J.: Towards inclusive tourism? Stakeholder collaboration in the development of accessible tourism. Tour. Geogr. **20**(4), 675–697 (2018)

Page, S., Yeoman, I.: How VisitScotland prepared for a flu pandemic. J. Bus. Contin. Emer. Plann. **1**(2), 167–182 (2006)

Pan, Z., Cheok, A.D., Yang, H., Zhu, J., Shi, J.: Virtual reality and mixed reality for virtual learning environments. Comput. Graph. **30**(1), 20–28 (2006)

Pantelidis, V.S.: Virtual reality in the classroom. Educ. Technol. **33**(4), 23–27 (1993)

Park, S.-M., Kim, Y.-G.: A metaverse: taxonomy, components, applications, and open challenges. IEEE Access **10**, 4209–4250 (2022)

Ramadhan, A., Suryodiningrat, S.P., Mahendra, I.: The fundamentals of metaverse: a review on types, components and opportunities. JIOS **47**(1), 153–165 (2023)

Rather, R.A.: Metaverse marketing and consumer research: theoretical framework and future research agenda in tourism and hospitality industry. Tour. Recreat. Res. **0**(0), 1–10 (2023)

Revfine. Metaverse tourism: overview, benefits, examples and more, 13 February 2022. https://www.Revfine.Com/Metaverse-Tourism/

Şahin, B., Güven, S.: Sağlık inanç modeli, turizm fobisi ve salgın hastalıklar. Güncel Turizm Araştırmaları Dergisi **6**(2), 25–43 (2022)

Saragih, R.E., Suyoto: Development of interactive mobile application with augmented reality for tourism sites in Batam, pp. 512–517. IEEE (2020)

Shubtsova, L.V., Kostromina, E.A., Chelyapina, O.I., Grigorieva, N.A., Trifonov, P.V.: Supporting the tourism industry in the context of the coronavirus pandemic and economic crisis: social tourism and public-private partnership. J. Environ. Manage. Tour. **11**(6), 1427–1434 (2020)

Suanpang, P., Niamsorn, C., Pothipassa, P., Chunhapataragul, T., Netwong, T., Jermsittiparsert, K.: Extensible metaverse implication for a smart tourism city. Sustainability **14**(21), 1–19 (2022)

Tayfun, A., Silik, C.E., Şimşek, E., Dülger, A.S.: Metaverse: Turizm için bir fırsat mı? yoksa bir tehdit mi? J. Tour. Gastron. Stud. **10**(2), 818–836 (2022)

Trunfio, M., Rossi, S.: Advances in metaverse investigation: streams of research and future agenda. Virtual Worlds **1**, 103–129 (2022)

Tsai, S.-P.: Investigating metaverse marketing for travel and tourism. J. Vacat Mark. **0**(0), 1–10 (2022)

Um, T., Kim, H., Kim, H., Lee, J., Koo, C., Chung, N.: Travel incheon as a metaverse: Smart tourism cities development case in Korea. In: Stienmetz, J.L., Ferrer-Rosell, B., Massimo, David (eds.) ENTER 2022, pp. 226–231. Springer, Cham (2022). https://doi.org/10.1007/978-3-030-94751-4_20

Wang, D., Yan, X., Zhou, Y.: Research on metaverse: Concept, development and standard system. Jn: 2nd International Conference on Electronics, Communications and Information Technology, pp. 983–991. IEEE, Sanya (2021)

Wei, D.: Gemiverse: the blockchain-based professional certification and tourism platform with its own ecosystem in the metaverse. Int. J. Geoherit. Parks **10**(2), 322–336 (2022)

Wei, W.: A buzzword, a phase or the next chapter for the Internet? The status and possibilities of the metaverse for tourism. J. Hospit. Tour. Insight **0**(0), 1–24 (2023)

Yang, Q., Zhao, Y., Huang, H., Xiong, Z., Kang, J., Zheng, Z.: Fusing blockchain and AI With metaverse: a survey. Comput. Soc. **3**, 122–136 (2022)

Yemenici, A.D.: Entrepreneurship in the world of metaverse: virtual or real? J. Metaverse **2**(2), 71–82 (2022)

Yuviler-Gavish, N., et al.: Evaluating virtual reality and augmented reality training for industrial maintenance and assembly tasks. Interact. Learn. Environ. **23**(6), 1–21 (2013)

Zaman, U., Koo, I., Abbasi, S., Raza, S.H., Qureshi, M.G.: Meet your digital twin in space? Profiling international expat's readiness for metaverse space travel, tech-savviness, COVID-19 travel anxiety, and travel fear of missing out. Sustainability **14**(11), 1–19 (2022)

Zheng, J., Chan, K., Gibson, I.: Virtual reality. IEEE Potential **17**(2), 20–23 (1998)

Zhong, L., Sun, S., Law, R., Li, X.: Tourism crisis management: evidence from COVID-19. Curr. Issue Tour. **24**(19), 2671–2682 (2021)

Transitioning the Tourism Industry from the Real World to the Metaverse

Neha Sharma[1](✉), Neetima Aggarwal[2], Mahesh Uniyal[3], Gul Erkol Bayram[4], and Vijay Prakash[5]

[1] Indian Institute of Management, Sirmaur, India
tourism.neha@gmail.com
[2] Jaypee Institute of Information Technology University, Noida, India
setneetima@gmail.com
[3] School of Hotel Management and Tourism, Dev Bhoomi, Uttarakhand University, Dehradun, India
mcuniyal76@gmail.com
[4] School of Tourism and Hospitality Management, Department of Tour Guiding, Sinop University, Sinop, Turkey
gulerkol@sinop.edu.tr
[5] Department of Tourism Studies, Dharmanand Uniyal Government Degree College, Narendranagar, Uttarakhand 249175, India
vijaybhatt66@gmail.com

Abstract. Metaverse exists. Various qualities, however, metaverse tourism is not yet recognized. It provides four metaverse tourist ideas. The primary metaverse tourist technology will enhance immersion. Pre-trip metaverse passengers can expect realistically. Third, examine metaverse travelers' multi-identities. Finally, creative economic metaverse tourism is new. Metaverse tourism ecology before, during, and after. Metaverse could change life. Cellphones and the Internet revolutionized travel. Tourism is theoretically linked to Metaverse dimensions. Tourists should expect metaverse evolution to change their experiences and disintegrate tourism. Rapid digital technology proliferation created the metaverse, a virtual shared environment with physical and digital worlds. The metaverse is prompting sectoral changes and new engagement channels. The tourism industry has changed. Virtual, augmented, and other immersive technologies allow tourism to be customized across borders. Tourism using metaverse and apps. Cultural exhibits, virtual city tours, historical reenactments, and immersive nature adventures are examples. Metaverse technology developers, tourist stakeholders, and legislators must collaborate. Technological advances are reshaping the metaverse, a virtual shared realm of physical and digital worlds. Some industries are testing it to disrupt practices and create new engagement channels. Tourism has altered drastically. According to polls, metaverse experiences attract travelers. The metaverse's principles and tourism effects show its revolutionary potential to change travelers' venue and experience interactions. VR, AR, and immersive tourism tailor international travel. Tourist case studies in industry and metaverse offer virtual city tours, historical reenactments, interactive cultural exhibits, and immersive nature activities. Technology developers, tourist stakeholders, and the government must work together on the metaverse. Regulation and ethics protect metaverse tourism.

M. Al-Emran et al. (Eds.): IMDC-IST 2024, LNNS 876, pp. 21–36, 2023.
https://doi.org/10.1007/978-3-031-51300-8_2

Keywords: Metaverse · Tourism · virtual reality · augmented reality

1 Introduction

The word "metaverse" is derived from the combination of the words "meta" and "universe." This sentence came from Snow Crash, 1992. The word "metaverse" refers to a type of three-dimensional virtual environment that places a strong emphasis on interpersonal connections (Gursoy et al., 2022; Hamid et al., 2021). In this environment, people can converse with one another either through their avatars or as genuine participants. A virtual world (or transcendence) that exists outside the realm of the physical world is sometimes referred to as the metaverse (Koo et al., 2023). Aristotle considered physics the fundamental law of nature and existence. In contrast, "metaphysics," meaning "after physics," discusses awareness and mind-matter interactions. VR and online second life appear to be the main issue. It is possible to break down new metaverse concepts into some individual components, such as augmented reality (AR), a digital second life, an immersive 3D environment, a computer platform, and avatars that engage in social interaction with other people. Tourism is full of articulate things, humans, avatars, interfaces, and networking capabilities. This is what is known as the metaverse tourism. Because of advances in technology, the tourist industry has begun implementing innovations to enhance the overall experience of its clients while simultaneously lowering its operational costs (Buhalis et al., 2023; Eneizan et al., 2019). In this context, a great number of academics have presented new ideas in the field of e-tourism, smart tourism (Go and Kang, 2023; Al-Abrrow et al., 2021). Scholars are studying third Internet wave, which brought digital twins and the metaverse, due to the growing importance of understanding technological revolution paradigms in travel (Um et al., 2022; Khaw et al., 2022).

Hospitality and tourism have noticed the metaverse because it inspires passengers to travel to a broad geographic areas beyond limits, fostering presence and social interaction (Tsai, 2022; Alsalem et al., 2022). For instance, Southwest Airlines is now offering its passengers a moving map application that can be used while they are in flight. Passengers will have the ability to use augmented reality (AR) and view short movies thanks to the map, provides a realistic virtual tour of Southwest destinations. Leven, a brand in the hospitality industry, is in the process of creating a metaverse virtual hotel in which individuals from all over the world will be able to connect in an exciting and engaging virtual hotel setting. Metaverse could be used to evaluate consumers in the future, there is no guarantee that this will occur (Monaco and Sacchi, 2023; Albahri et al., 2021). It is still something that has to be developed to construct a comprehensive tourist ecosystem in the metaverse and gain a knowledge of the benefits of metaverse tourism (Buhalis and Karatay, 2022; Fadhil et al., 2021).

Only in the most recent decades has the phenomena of tourism been given attention as a subject of scientific investigation, with the acknowledgment of its scale and the potential implications it may have on society (Buhalis et al., 2022; Alnoor et al., 2022). For a considerable amount of time, it was thought of as a unified and well-defined endeavor that was predominately predicated on the idea that participants physically move to pre-determined locations using appropriate modes of transport to spend the allotted amount of time there. This understanding persisted for a considerable length of time (Suanpang et al., 2022; AL-Fatlawey et al., 2021). An assumption like this one established the foundation for a modern interpretation of the tourism phenomena.

It is common knowledge that new technologies can radically alter life. For the past few decades, ICTs have revolutionized communication altered the travel and tourism sector (Volchek and Brysch, 2023; Gatea and Marina, 2016). Tourism service providers, tour operators, and middlemen are reorganizing their production processes for travel and leisure items while also developing new models for addressing clients. These models include investments in new sales channels. Because of the rise of digital technology, new forms of tourism have evolved, including VR museum, amusement park, and escape room tours (Yang and Wang, 2023; Manhal et al., 2023). The Fourth Industrial Revolution has brought about structural changes in the market players themselves, particularly supply and demand. It has expanded tourist experience enhancement potential and has posed a challenge to the competitiveness of businesses through the implementation of innovations.

2 Literature Review

Tourism and economic policy uncertainty (EPU) have garnered policymaker and researcher interest. Tourism, the "smokeless industry," is a fast-growing sector that drives socio-economic growth. In 2017, 37.7 million tourists visited the UK, World Tourism Organization (2018). Travel and tourism account for almost 3% of GDP and 5% of jobs. ITM produces money from transportation, lodging, food, and beverage, and cultural attractions. Terrorist attacks and the Iraq conflict caused instability, which impacted ITM between 2001–2003. The UK's ITM dropped due to uncertainties in the US, Germany, Japan, and its biggest tourist markets. EPU dropped in 2005 and the tourism industry resumed its previous pace, indicating a sharp decline. According to the World Tourism Organization, the UK received £16 billion in ITM revenues in 2006, ranking sixth in travel sector revenues. From 2000 to 2008, ITM experienced significant expansion. This period was marked by robust economic growth and a decline in the EPU in the UK, promoting tourism.

It examines the correlation in only one direction, emphasizing the unidirectional effect of EPU on tourism. Consequently, this study endeavors to explicate the interdependence of EPU and ITM. The findings indicate that EPU has a deleterious impact on ITM in the subsamples. ITM has a positive effect on EPU, however. Financial and economic events that led to an increase in EPU and a decrease in ITM can be considered causal. In addition, the results indicate that EPU plays a considerable role in ITM, which is supported by the gravity model, which indicates that on multiple occasions, ITM decreased due to EPU. Second, this study shows a correlation between EPU and ITM in the UK, a major tourist destination, and counts the exchange rate (EXG) as a control variable. The EXG substantially increases short-term uncertainty, affecting travelers' plans. EPU negatively affects ITM in subsamples. EPU benefits from ITM. Eventual financial and economic factors increased EPU and decreased ITM. The gravity model shows that EPU decreases ITM numerous times, supporting the results. This study shows a correlation between EPU and ITM in the UK, a popular tourist destination, and calculates the exchange rate, suggesting that their dynamic relationship is unstable. It considers the time-varying properties of the time series. EPU and UK tourism are assessed in the face of structural changes using the rolling window. Multiple subsamples suggest Khan et al. 3 bidirectional causalities between EPU and ITM. The gravity model suggests that a greater EPU hurts ITM in the UK (Wei, 2022; Atiyah and Zaidan, 2022).

The IMF was founded to protect the global financial system. The Fund represents the advanced industrial countries that provide most of its resources, which have a strong interest in financial stability and conservative fiscal management, privatization, and trade liberalization in the developing world.

2.1 Metaverse: The Intersection of "Meta" and "Universe

"The words "meta" and "universe" are where the term "metaverse" originates (Martins et al., 2022). The Metaverse is a collective virtual shared place that was established by the confluence of virtually enhanced physical reality and physically persistent virtual space. It includes the sum of all virtual worlds, augmented reality, and the internet (Chen et al., 2023; Atiyah, 2023). Neal Stephenson, a writer of science fiction, originally used the term "metaverse" in the novel "Snow Crash," which he published in 1992. The "Snow Crash" metaverse is an extremely well-liked virtual environment that users who are armed with augmented reality equipment can explore (Wei, 2023). Realistic virtual worlds are simulations of the physical world in which humans are represented by computer avatars. The decisions that are made and the acts that are taken by society have a direct impact on the virtual world, which is always expanding and changing. With the use of augmented and mixed reality, people will be able to enter the metaverse wholly and virtually or interact with elements of it in their actual location. This will be possible for them to do either. A virtual digital environment can be created with the help of Metaverse, and this world can then be connected with the real world (Lin et al., 2023). Mark Zuckerberg announced that Facebook would be rebranded as Meta in October 2021, and the company would also make substantial expenditures in the metaverse (Gomes and Araujo, 2012; Atiyah, 2023).

It has been suggested that the metaverse is one of the transformative technologies that could have a game-changing impact on the way people live their lives. It goes beyond augmented reality (AR) and virtual reality (VR) and provides the game-changing experience of being in a virtual world with large-scale economic, social, and cultural connections (Wong et al., 2023). As a result, Disney is working towards combining the Disneyland and Disney World theme parks with real-time virtual concerts, shops for virtual avatars, and augmented reality location-based personalized dialogue with Disney characters (Jafar and Ahmad, 2023). Large technological corporations, venture capital firms, private equity firms, startup enterprises, and well-known brands are all interested in capitalizing on the opportunities that the Metaverse presents. More than double the $57 billion that was generated from investments made by these companies in 2021, these businesses generated more than $120 billion in the first five months of 2022 (Ioannidi and Kontis, 2023). In the presentation, you should emphasize the fact that the Metaverse will have a revolutionary impact on tourism. On the other hand, the extent of the potential shifts that could occur in tourism as a result of the influence of the Metaverse is not yet fully understood (Table 1).

Table 1. Conceptualization of Metaverse Tourism.

	E Tourism	Smart Tourism	Metaverse Tourism
Sphere	Digital	Connecting the digital world with the real world	The transition from the digital to the physical is completely seamless
Core technology	Web search engine, also HTML	Sensors and cellphones, large amounts of data, artificial intelligence for automobiles, robots for service	BMI, HMD, 3D brainstorming tools, facial expression tools, rendering technologies, blockchain, bitcoin, and non-fungible tokens are some of the terms that have been thrown around
Role of technology	Looking, making reservations, and analyzing	Getting around and making mobility better	Simulation, a feeling of being, the appearance of an avatar, and the recalling of memories
Phase	Pre- and post-trip	During trip	Pre-, during, and post-trip
Economic system	Transaction economy	Platform economy	Metaverse creative economy
Tourist identity	Human	Human	Avatar and human
Information cues	Text and image	Text, image, and video (e.g., YouTube	Text, images, and videos, along with one or more of the five bodily senses (such as hearing, seeing, or touching)
Tourist experience	Information search and booking	Cocreation with technology mediation	Interactions of a parasocial nature between visitors and their avatars
Outcomes	Competition in pricing, delivery, and logistics, as well as improvements in product and service	Recommendation, activities, and interactions inside social networks, personalization, making decisions based on online reviews, and peer-to-peer service modification	Image of oneself, harmony with people in a virtual environment, the production of creative goods and objects, and the establishment and maintenance of a connection between the metaverse and the physical world are some of the topics that will be covered

2.2 Metaverse Tourism Ecosystem

The expansion of digital capacities and capabilities has led to the emergence of novel business models in the ecosystems of 3D web computing. The current study suggests a metaverse tourism ecosystem by utilizing the conceptual idea. This ecosystem would consist of four different types of business models (for example, augmented reality, lifelogging, mirror worlds and virtual worlds), two sides (a traveler and a supplier), and two complementary worlds (for example, the real world and the virtual world). There are two technology options available for doing different kinds of business. The term "augmentation" refers to "technologies that add new capabilities to existing real systems via a physical layer. "Technologies that provide simulated worlds as the locus for interaction" is one definition of the term "simulation." Because of this, provide mirror worlds "informationally enhanced virtual models and reflections of the physical world," as well as lifelogging, which is defined as "user life histories, observation, communication, and behavior modeling of everyday experiences". Figure 1 illustrates the classification of the metaverse into its various types according to horizontally (intimate (world-focused) and exterior (identity focus) vertical augmentation and simulation (Chew et al., 2023; Abbas et al., 2023).

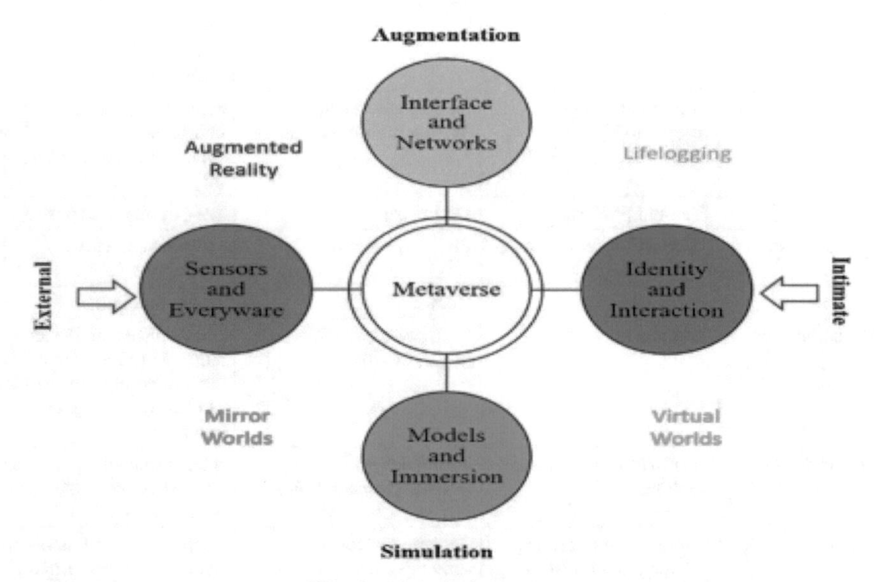

Fig. 1. Metaverse typology

An immersive 3D virtual world called a "metaverse tourism ecosystem" supports travelers' social interactions with each other and travel agents and offers opportunities to create a new method for cannibalizing services and inspire a creative economy by connecting virtually and reconnecting physically (Dutta et al., 2023).

2.3 The Contemporary Phenomenon of Tourism

Tourism is a unique category of human activity, which has received a lot of attention in recent years. (Murti et al., 2023) outlined the five "elements" that contribute to the formation of a tourist system that is both functional and spatial: tourists, locations that generate demand, regions that are destinations, an industry that caters to tourists, and transportation routes. (Jo, 2023) It explains that tourism is fundamentally made up of two subsystems, which are referred to as "tourism-subject" (also known as tourists) and "tourism-object" (also known as tourist places, tourism businesses, and tourism organizations). The modern definition of tourism offered by the UNWTO (Fazio et al., 2023) emphasizes the individuals who travel, the reasons for traveling, locations that are outside of the visitors' usual setting and process of transforming a regular area into a tourist destination as the fundamental components of the tourism phenomenon. People who produce tourism demand, locations that are capable of satisfying this want, an ecology that offers basic services, and most importantly, a tourist retreat that addresses tourism demand are the factors that decide whether or not tourism exists as a phenomenon on its own.

Tourist Needs and the Purpose of Travel. According to UNWTO (Özdemir and Şahin, 2023), people travel for a variety of reasons, including "business, leisure, or other personal" reasons. Presents a more diverse classification of tourist needs (pleasure, commerce, personal quest, nature, and human endeavor) that should be addressed to introduce a taxonomy of tourist products. These needs include pleasure, business, nature, and human endeavor. In addition, verse stresses the importance of socio-psychological and cultural diversification as an escape from the monotony of daily life. Demand for tourism and the ability of tourist products to meet that demand have a significant impact on the industry (Rather, 2023).

Tourist Destination and Place Concept. The need to satiate one's requirements compels people to go to greater lengths than is typical for them. The place is described as any destination that is outside of the tourist's typical environment (Hui et al., 2023). This definition applies to the place regardless of its nature, size, popularity, etc. It is common practice to differentiate between "usual" environments and tourist environments based on factors such as the frequency of visits or the length of stays, as well as the physical distance traveled (Baker et al., 2023). Stress the fact that being "outside of a usual environment" could entail being in a virtual location or a mental state in which one is at ease with oneself. The idea of a tourist destination does not simply refer to a specific geographical location; rather, it can refer to any setting that is distinctive in comparison to the typical one.

Tourism Ecosystem. Tourism is a phenomenon that is characterized by the economic activities of demand generators who are away from home and by the supply of lodging and other services to meet this demand. Tourism is defined as a phenomenon that occurs when demand generators away from home do these. According to McIntosh and Goeldner (Huang et al., 2023), tourism is defined as a collection of distinct tourism activities along with the services (such as lodging, dining, transportation, shopping, and entertainment) and commercial enterprises that are associated with them. Tourism, which is one of the major sectors in the world, has the potential to be both beneficial and damaging to the surrounding socioeconomic and cultural environment. The long-term growth and sustainability of tourist sites, national economies, and entire regions must have a comprehensive grasp of tourism as a phenomenon (Zhang and Quoquab, 2023).

Escape from the Usual Environment to Become a Tourist. The differentiators are spatial and temporal to assess the effects of tourism. A visitor is distinguished from a tourist by the requirement that the latter spend at least one night away from his or her usual residence (Corne et al., 2023). One has been chosen as the maximum threshold that is recommended for use in differentiating tourists from permanent residents. Escape is related to the physical state, mind, and experience, such as relaxation, nature, and gastronomy, and so on. Tourism is widely referred to as a liminal activity, which is described as a "transition from known to unknown" and is related to the change in human behavior from regular routines to frequently atypical choices. The criteria for switching from normal behavior to that of a tourist and back again have not been fully investigated (Go et al., 2023).

The tourism phenomenon is not a fixed phenomenon. Throughout its history, various social, economic, political, and other elements have served as the impetus for successive stages of its development. They have also altered people's conceptions of what a traveler is, what the goal of travel may be, and how it can influence the process of sustainable development (Um et al., 2022). The adaptability of tourism is directly proportional to the phenomenon's level of complexity as well as its dimensions. It is anticipated that tourism footprints, political significance, and transformative technology will be the primary drivers of future turning points in the growth of tourism (Tsai, 2022). The Fourth Industrial Revolution made possible whole original kinds of vacationing and pastimes to enjoy in one's spare time. Because of this, going back in time to the Middle Ages through video games is now possible. The limits of space and time are currently being pushed to their limits. The conventional conception of tourism may end up being insufficiently broad to accommodate the emerging reality. On the other hand, the connection of the tourism phenomenon with the Metaverse is not yet well comprehended. Both ideas involve humans in some capacity, be it as users or tourists; they can address either actual or virtual locations; they define otherness through experiences; and, ultimately, they both require an ecosystem to work. To comprehend the future of tourism and how it will affect society, it is vital to consider the possibility of these components coming into alignment with one another (Table 2).

ARIMA is popular for time series data analysis. ARIMA includes:

- The Autoregressive (AR) section explains linear regression between current and past data.
- Moving Average (MA) explains linear regression between current data and prior forecast inaccuracies due to white noise. AR/MA only works for stationary data.

ARIMA, the "Integrated" part, improves AR and MA models for non-stationary data. A batch learning method, the ELM model trains and works with all available data. The model cannot learn from new data while running. Therefore, a research study presented the Online Sequential Extreme Learning Machine (OS-ELM) to improve the ELM model for incremental learning (Monaco and Sacchi, 2022) (Table 3).

2.4 Managerial Implications of Transitioning the Tourism Industry to The Metaverse

- **Market Research and Understanding User Behavior:** Conduct extensive market research to understand the target audience's preferences and expectations within the metaverse. Analyze user behavior and trends in the metaverse to adapt tourism offerings accordingly.
- **Digital Presence and Branding:** Establish a strong digital presence and a recognizable brand in the metaverse. Maintain consistency between real-world and metaverse branding to build trust and recognition.

Table 2. Comparison of existing literature Review.

Authors (Year)	Methods	Focus	Key Findings/Results
Yeoman et al. (2019) [43]	Historical analysis, literature review	Evolution of tourism from 1946 to 2020	Identified historical turning points in the evolution of tourism (mobility, Fordism, mass tourism) and future turning points (political significance, footprints, transformative technology). The conclusion is that discussions about tourism from the past continue to be relevant and essential today and in the future
Cunha et al. (2012) [42]	Analysis of international organizations and academic literature	Defining the concept of tourism	Examined inadequacies of extant tourism definitions and sought to contribute to the development of a scientifically accepted definition
Li, X.R et al. (2019) [33]	Comparative analysis	Comparison of first-time and repeat tourists	Identified demographic, typographic, and behavioral differences between first-time visitors and recurrent tourists. Tourism and travel dominated the interests of first-time visitors, while recreation and activity dominated the interests of frequent guests
Smart et al. (2007) [27]	Literature review	Applications of the metaverse in education	Examined the potential benefits and obstacles of incorporating the metaverse into educational settings, focusing on its interactivity and the need for structural adaptation
Koo, Chulmo, et al. (2020) [26]	Conceptual analysis	Metaverse tourism hypotheses	Four hypotheses regarding metaverse tourism are proposed, including immersive experiences, pre-trip adjustment of expectations, multi-identity profiles, and a new paradigm for the creative economy. Proposed investigation of metaverse tourism during travel phases

(*continued*)

Table 2. (*continued*)

Authors (Year)	Methods	Focus	Key Findings/Results
Buhalis et al. (2022) **[25]**	Semi-structured interviews, qualitative research	Role of Mixed Reality (MR) in Tourism for Generation Z	Analyzed the perspective of Generation Z on immersive technology (MR) for cultural heritage experiences. Technology-enabled co-creation of transformative experiences was highlighted
Davis et al. (2009) [2]	Conceptual model, literature review	Metaverses for virtual collaboration	Presented a conceptual paradigm for the study of metaverses with a focus on five fundamental constructs. Highlighted metaverse technology's potential to facilitate more robust and engaging virtual collaboration

- **Content Creation and Virtual Experiences:** Invest in creating high-quality virtual experiences that resonate with users. Develop immersive content that showcases destinations, hotels, attractions, and activities within the metaverse.
- **Technology Integration:** Invest in cutting-edge technology for virtual reality (VR) and augmented reality (AR) to enhance user experiences. Ensure seamless integration of technology platforms for bookings, reservations, and payments.
- **User Engagement and Community Building:** Foster a sense of community within the metaverse by organizing events, meetups, and social interactions. Encourage user-generated content and interactions to build a vibrant virtual tourism ecosystem.
- **Security and Privacy:** Implement robust security measures to protect user data and privacy within the metaverse. Comply with relevant data protection regulations to maintain user trust
- **Monetization Strategies:** Develop sustainable monetization models, such as virtual ticketing, in-app purchases, and virtual goods sales. Explore partnerships and collaborations to expand revenue streams.
- **Sustainability and Eco-Friendly Initiatives:** Emphasize sustainability and eco-friendly practices within the metaverse, mirroring real-world efforts. Highlight destinations and experiences that promote sustainable tourism.
- **Regulatory Compliance:** Stay informed about evolving regulations and policies related to virtual tourism and the metaverse. Comply with legal requirements to avoid potential legal issues.
- **Feedback and Iteration:** Continuously gather user feedback and adapt offerings based on user preferences and changing technology. Be agile and ready to pivot strategies as the metaverse evolves.
- **Cross-Industry Collaborations:** Collaborate with other industries, such as tech companies, entertainment, and gaming, to create innovative metaverse experiences. Explore synergies that can enhance the tourism offering within the metaverse.

Table 3. Comparison of the ARIMA model and Os- ELM.

Aspect	ARIMA	Extreme Learning Machine (ELM)
Type	Statistical time series forecasting model	Machine learning algorithm
Model Complexity	Typically requires manual parameter tuning	Simpler, fewer hyperparameters
Stationarity	Assumes stationarity or requires differencing	Can handle non-stationary data
Seasonality	Handles seasonality with seasonal differencing	May require additional techniques
Interpretability	Provides insight into the time series components	Less transparent, more of a black box
Training Speed	Slower due to iterative parameter estimation	Faster training due to random initialization
Handling Outliers	Sensitive to outliers, and may require data preprocessing	Robust to outliers if properly designed
Handling Missing Data	Can be challenging with missing data	Handles missing data better
Model Flexibility	Suitable for linear and non-linear patterns	Capable of capturing complex non-linear patterns
Forecast Accuracy	Can perform well when assumptions are met	Performance may vary based on data and model design
Software and Libraries	Available in various statistical packages (e.g., R)	Implementations are available in Python and other languages

- **Marketing and Promotion:** Develop a marketing strategy tailored to the metaverse, leveraging social media, influencers, and virtual advertising. Engage in cross-promotion with other metaverse-based businesses to expand reach.
- **Risk Management:** Assess potential risks associated with virtual tourism, such as technical glitches, cyberattacks, or negative user experiences. Develop risk mitigation strategies to address these challenges.

3 Discussion

Rapid technological advancement has continually reshaped industries and human experiences, and the emergence of the metaverse is poised to revolutionize how we perceive and interact with the world. This discussion explores the implications, challenges, and opportunities of transitioning the tourism industry from the physical domain to the metaverse.

1. **Embracing the Metaverse: A Paradigm Shift in Tourism.** The term "metaverse," popularized by "Snow Crash," Neal Stephenson describes a virtual reality. Environment that transcends the physical world and emphasizes interpersonal connections and immersive experiences. It is a space in which people interact with each other and digital elements using avatars, thereby obscuring the line between reality and virtuality. By combining augmented and mixed reality technologies, individuals can enter the metaverse from any location and experience new dimensions that transcend physical limitations.
2. **The Metaverse Tourism Ecosystem: Augmented Reality, Lifelogging, and More.** The transition to metaverse tourism introduces a novel ecosystem consisting of a variety of business models, stakeholders, and technological components. This ecosystem is comprised of various components, including augmented reality (AR), lifelogging, mirror worlds, and virtual worlds. Augmentation technologies augment real-world systems with additional capabilities, whereas simulations provide informationally enriched models of the physical world. These technologies converge to create a dynamic environment that offers distinct and immersive experiences to travelers.
3. **Implications for the Hospitality and Tourism Industries.** The prospective impact of the metaverse on the hospitality and tourism industries is extensive. In a post-pandemic world, its function in fostering a presence and social interaction is particularly noteworthy. Companies are investigating metaverse applications such as virtual travel experiences and interactive virtual hotels, as demonstrated by Southwest Airlines and Leven. However, obstacles remain, such as the need to develop exhaustive consumer evaluation tools and establish seamless connections between metaverse and reality.
4. **Metaverse Tourism: A New Frontier of Experiences.** Metaverse tourism promises to improve how tourists seek information, make reservations, and interact with locations. This shift from text- and image-based information indicators to multisensory interactions may reshape the tourist experience. The capacity of the metaverse to generate competition, personalization, and peer-to-peer service modification has the potential to revolutionize how travelers make decisions and interact with their surroundings.
5. **The Metaverse as a Catalyst for Technological Innovation.** The Fourth Industrial Revolution has enabled the emergence of the metaverse and its incorporation into the tourism industry. As technology giants and innovative startups invest significantly in the development of the metaverse, its potential impact on the economic landscape of the industry is significant. Beyond augmented and virtual reality, the metaverse's transformative power presents a unique opportunity for businesses to develop innovative products and services that appeal to a tech-savvy and digitally engaged tourist population.
6. **Challenges.** Despite the metaverse's allure, there are still several obstacles to overcome. The extent to which the metaverse will reshape tourism and the full scope of its potential consequences are not yet completely understood. Ensuring a seamless integration between the digital and physical realms, addressing ethical concerns related to privacy and data utilization, and managing potential disparities in access to metaverse experiences are crucial issues that must be addressed.

7. **Envisioning the Future of Metaverse Tourism.** The tourism industry's transition from reality to metaverse is a revolutionary shift that offers unprecedented opportunities to both tourists and businesses. The ability of the metaverse to generate immersive and interconnected experiences has the potential to revolutionize how visitors interact with destinations and services. To completely realize the potential of metaverse tourism, however, technological advances, ethical considerations, and the changing needs of a dynamic and diverse tourist population must be carefully considered.

4 Conclusion

Potential metaverse tourism influence on travel. Metaverse tourism ecosystems must serve multiuser virtual worlds and a fast-growing digital environment with real-world social, economic, tourism, and political activities. Metaverse tourism should provide a platform that seamlessly combines broadband and Wi-Fi infrastructure to run 3D rendering technologies at any time. Pre--, during-, and post-travel metaverse tourism experiences are examined to provide precise concepts and solutions. Social, business, events, attractions, places, and destinations were tourism. Travel and Metaverse. Strong connections exist between the two ideas. Studies show tourism and metaverse growth interact. Tourism companies must consider numerous ecodevelopment possibilities. Tourism could be transformed by metaverse technology. Technology continuously alters our views. A metaverse drives this technology revolution beyond physical limitations. The tourism industry's metaverse growth offers intriguing potential and puzzling obstacles. Augmented, virtual, and immersive metaverse experiences may transform travel, recreation, and exploration. A new interactional dimension merges fact and fiction. Tourists may easily travel digital landscapes, interact with other explorers via avatars, and create metaverse memories that mix the physical and virtual worlds. After worldwide disasters like COVID-19, this paradigm shift allowed remote experiences and social relationships.

Tech metaverse alters industry. The innovative tourism company uses the metaverse to improve customer experiences, save money, and be creative. From immersive in-flight experiences to virtual hotels that join people globally, the metaverse is altering tourism by offering new ways to explore, participate in, and visualize destinations. As with any technical breakthrough, the metaverse transition has problems. The complex relationship between real and virtual worlds raises authenticity, privacy, and cultural heritage issues. Virtual experiences are useful, but connections are priceless. Reevaluating commercial and regulatory tactics is important for responsible metaverse expansion. The complicated tourist network must be considered while envisioning metaverse tourism. Tourism must adapt to different travel incentives and complicated site, company, and service ecology. In the metaverse, these aspects can combine, allowing travellers new opportunities to meet their needs, explore new destinations, and have fantastic experiences. Fourth Industrial Revolution metaverses may change tourism. Tourism may change with metaverse inclusion. This future requires technological innovation and a profound understanding of tourism's complex dynamics as a transformative human activity. Finally, the metaverse helps tourism rethink and improve global exploration. The metaverse combines digital and physical domains to generate cross-boundary experiences. Tech and curiosity will make tourism's future fascinating and surprising.

References

Abbas, S., et al.: Antecedents of trustworthiness of social commerce platforms: a case of rural communities using multi group SEM & MCDM methods. Electron. Commer. Res. Appl. **62**, 101322 (2023)

Al-Abrrow, H., Fayez, A.S., Abdullah, H., Khaw, K.W., Alnoor, A., Rexhepi, G.: Effect of open-mindedness and humble behavior on innovation: mediator role of learning. Int. J. Emerg. Markets (2021).

Albahri, A.S., et al.: Based on the multi-assessment model: towards a new context of combining the artificial neural network and structural equation modelling: a review. Chaos Solitons Fractals **153**, 111445 (2021)

AL-Fatlawey, M.H., Brias, A.K., Atiyah, A.G.: The role of strategic behavior in achievement the organizational excellence "Analytical research of the manager's views of ur state company at Thi-Qar governorate". J. Adm. Econ. **10**(37), 48–68 (2021).

Alnoor, A., et al.: How positive and negative electronic word of mouth (eWOM) affects customers' intention to use social commerce? A dual-stage multi group-SEM and ANN analysis. Int. J. Hum. Comput. Interact. 1–30 (2022).

Alsalem, M.A., et al.: Rise of multiattribute decision-making in combating COVID-19: a systematic review of the state-of-the-art literature. Int. J. Intell. Syst. **37**(6), 3514–3624 (2022)

Atiyah, A.G.: The effect of the dimensions of strategic change on organizational performance level. PalArch's J. Archaeol. Egypt/Egyptology **17**(8), 1269–1282 (2020)

Atiyah, A. G. Strategic Network and Psychological Contract Breach: The Mediating Effect of Role Ambiguity (2023).

Atiyah, A.G., Zaidan, R.A.: Barriers to using social commerce. In: Alnoor, A., Wah, K.K., Hassan, A. (eds) Artificial Neural Networks and Structural Equation Modeling, pp. 115–130. Springer, Singapore (2022). https://doi.org/10.1007/978-981-19-6509-8_7

Baker, J., Nam, K., Dutt, C.S.: A user experience perspective on heritage tourism in the metaverse: empirical evidence and design dilemmas for VR. Inf. Technol. Tourism **25**, 1–42 (2023)

Buhalis, D., Karatay, N.: Mixed reality (MR) for generation Z in cultural heritage tourism towards metaverse. In: Stienmetz, J.L., Ferrer-Rosell, B., Massimo, D. (eds.) ENTER 2022, pp. 16–27. Springer, Cham (2022). https://doi.org/10.1007/978-3-030-94751-4_2

Buhalis, D., Leung, D., Lin, M.: Metaverse as a disruptive technology revolutionising tourism management and marketing. Tour. Manage. **97**, 104724 (2023)

Buhalis, D., Lin, M.S., Leung, D.: Metaverse as a driver for customer experience and value co-creation: implications for hospitality and tourism management and marketing. Int. J. Contemp. Hosp. Manag. **35**(2), 701–716 (2022)

Chen, S., Chan, I.C.C., Xu, S., Law, R., Zhang, M.: Metaverse in tourism: drivers and hindrances from stakeholders' perspective. J. Travel Tour. Mark. **40**(2), 169–184 (2023)

Chew, X., Khaw, K.W., Alnoor, A., Ferasso, M., Al Halbusi, H., Muhsen, Y.R.: Circular economy of medical waste: novel intelligent medical waste management framework based on extension linear Diophantine fuzzy FDOSM and neural network approach. Environ. Sci. Pollut. Res. **30**, 60473–60499 (2023)

Corne, A., Massot, V., Merasli, S.: The determinants of the adoption of blockchain technology in the tourism sector and metaverse perspectives. Inf. Technol. Tourism **25**, 1–29 (2023)

Dutta, D., Srivastava, Y., Singh, E.: Metaverse in the tourism sector for talent management: a technology in practice lens. Inf. Technol. Tourism **25**, 1–35 (2023)

Eneizan, B., Mohammed, A.G., Alnoor, A., Alabboodi, A.S., Enaizan, O.: Customer acceptance of mobile marketing in Jordan: an extended UTAUT2 model with trust and risk factors. Int. J. Eng. Bus. Manage. **11**, 1847979019889484 (2019)

Fadhil, S.S., Ismail, R., Alnoor, A.: The influence of soft skills on employability: a case study on technology industry sector in Malaysia. Interdiscip. J. Inf. Knowl. Manag. **16**, 255 (2021)

Fazio, G., Fricano, S., Iannolino, S., Pirrone, C.: Metaverse and tourism development: issues and opportunities in stakeholders' perception. Inf. Technol. Tourism **25**, 1–22 (2023)

Gatea, A.A., Marina, V.: Higher education funding in Iraq in terms of the experience of particular developed countries. Int. J. Adv. Stud. **6**(1), 8–17 (2016)

Go, H., Kang, M.: Metaverse tourism for sustainable tourism development: tourism Agenda 2030. Tourism Rev. **78**(2), 381–394 (2023)

Gomes, D.A., Araujo, M.C.B.: Tourism online: a metaverse offers study. Estudios y Perspectivas en Turismo **21**(4), 876–903 (2012)

Gursoy, D., Malodia, S., Dhir, A.: The metaverse in the hospitality and tourism industry: an overview of current trends and future research directions. J. Hosp. Market. Manag. **31**(5), 527–534 (2022)

Hamid, R.A., et al.: How smart is e-tourism? A systematic review of smart tourism recommendation system applying data management. Comput. Sci. Rev. **39**, 100337 (2021)

Huang, X.T., Wang, J., Wang, Z., Wang, L., Cheng, C.: Experimental study on the influence of virtual tourism spatial situation on the tourists' temperature comfort in the context of metaverse. Front. Psychol. **13**, 1062876 (2023)

Hui, X., Raza, S.H., Khan, S.W., Zaman, U., Ogadimma, E.C.: Exploring regenerative tourism using media richness theory: emerging role of immersive journalism, metaverse-based promotion, eco-literacy, and pro-environmental behavior. Sustainability **15**(6), 5046 (2023)

Ioannidis, S., Kontis, A.P.: Metaverse for tourists and tourism destinations. Inf. Technol. Tourism **25**, 1–24 (2023)

Jafar, R.M.S., Ahmad, W.: Tourist loyalty in the metaverse: the role of immersive tourism experience and cognitive perceptions. Tourism Review (2023)

Jo, H.: Tourism in the digital frontier: A study on user continuance intention in the metaverse. Information Technology & Tourism **25**, 307–330 (2023)

Khaw, K.W., et al.: Modelling and evaluating trust in mobile commerce: a hybrid three stage Fuzzy Delphi, structural equation modeling, and neural network approach. Int. J. Hum. Comput. Interact. **38**(16), 1529–1545 (2022)

Koo, C., Kwon, J., Chung, N., Kim, J.: Metaverse tourism: conceptual framework and research propositions. Curr. Issue Tour. **26**(20), 3268–3274 (2023)

Lin, K.J., Ye, H., Law, R.: Understanding the development of blockchain-empowered metaverse tourism: an institutional perspective. Inf. Technol. Tourism, 1–19 (2023).

Manhal, M., Al-khalidi, A., Hamad, Z.: Strategic network: managerial myopia point of view. Manage. Sci. Lett. **13**(3), 211–218 (2023)

Martins, D., Oliveira, L., Amaro, A.C.: From co-design to the construction of a metaverse for the promotion of cultural heritage and tourism: the case of Amiais. Procedia Comput. Sci. **204**, 261–266 (2022)

Monaco, S., Sacchi, G.: Travelling the metaverse: potential benefits and main challenges for tourism sectors and research applications. Sustainability **15**(4), 3348 (2023)

Murti, K.G.K., Darma, G.S., Mahyuni, L.P., Gorda, A.N.E.S.: Immersive experience in the metaverse: implications for tourism and business. Int. J. Appl. Bus. Res. 187–207 (2023).

Özdemir Uçgun, G., Şahin, S.Z.: How does Metaverse affect the tourism industry? Current practices and future forecasts. Curr. Issues in Tourism 1–15 (2023).

Rather, R.A.: Metaverse marketing and consumer research: theoretical framework and future research agenda in tourism and hospitality industry. Tourism Recreation Res. 1–9 (2023).

Suanpang, P., Niamsorn, C., Pothipassa, P., Chunhapataragul, T., Netwong, T., Jermsittiparsert, K.: Extensible metaverse implication for a smart tourism city. Sustainability **14**(21), 14027 (2022)

Tsai, S.P.: Investigating metaverse marketing for travel and tourism. J. Vacation Mark. 13567667221145715 (2022).

Um, T., Kim, H., Kim, H., Lee, J., Koo, C., Chung, N.: Travel incheon as a metaverse: smart tourism cities development case in Korea. In: Stienmetz, J.L., Ferrer-Rosell, B., Massimo, D. (eds.) Information and Communication Technologies in Tourism 2022: Proceedings of the ENTER 2022 eTourism Conference, January 11-14, 2022, pp. 226–231. Springer International Publishing, Cham (2022). https://doi.org/10.1007/978-3-030-94751-4_20

Volchek, K., Brysch, A.: Metaverse and tourism: from a new niche to a transformation. In: Ferrer-Rosell, B., Massimo, D., Berezina, K. (eds.) Information and Communication Technologies in Tourism 2023: Proceedings of the ENTER 2023 eTourism Conference, January 18-20, 2023, pp. 300–311. Springer Nature Switzerland, Cham (2023). https://doi.org/10.1007/978-3-031-25752-0_32

Wei, D.: Gemiverse: the blockchain-based professional certification and tourism platform with its own ecosystem in the metaverse. Int. J. Geoheritage Parks **10**(2), 322–336 (2022)

Wei, W.: A buzzword, a phase or the next chapter for the internet? The status and possibilities of the metaverse for tourism. J. Hospitality Tourism Insights. (2023)

Wong, L.W., Tan, G.W.H., Ooi, K.B., Dwivedi, Y.K.: Metaverse in hospitality and tourism: a critical reflection. Int. J. Contemp. Hospitality Manage. (2023).

Yang, F.X., Wang, Y.: Rethinking metaverse tourism: a taxonomy and an Agenda for future research. J. Hospitality Tourism Res. 10963480231163509 (2023).

Zhang, J., Quoquab, F.: Metaverse in the urban destinations in China: some insights for the tourism players. Int. J. Tourism Cities **9**(4), 1016–1024 (2023). https://doi.org/10.1108/IJTC-04-2023-0062

The Impact of Management Information Systems on International Human Resource Management: Moderating Role of Metaverse Culture

Hiba Yousif Al-Musawi[1] and Marcos Ferasso[2,3](✉)

[1] Business Administration Department, College of Administration and Economics, University of Basrah, Basrah, Iraq

[2] Escola de Ciências Económicas e das Organizações, Universidade Lusófona, Campo Grande, 376, 1749-024 Lisboa, Portugal

[3] Grupo de Investigación de Estudios Organizacionales Sostenibles, Universidad Autónoma de Chile, Santiago, Chile

marcos.ferasso@uautonoma.cl

Abstract. Many business models have benefited from metaverse applications. Virtual reality applications have facilitated human resources management in many companies. To this end, this study aims to investigate the impact of the information management system on international human resources management through the interactive role of metaverse culture. This study targeted 200 managers in international oil companies in southern Iraq. The results revealed an influential relationship between the information management system and international human resources management. The strength of the relationship between the information system and human resources management also increased when the metaverse culture was high.

Keywords: Metaverse Culture · Virtual reality · Human resource management · Information system

1 Introduction

As a result of developments in large companies in terms of size and stature, their need to obtain different human resources from diverse environments and diverse cultures has increased. Thus, the trends of successful companies today have become an international orientation after they were local. Therefore, in order to achieve its goal, it had to use global digital systems in order to facilitate the process of attracting various human resources, and thus these companies will obtain increasing overlap and international,

M. Al-Emran et al. (Eds.): IMDC-IST 2024, LNNS 876, pp. 37–53, 2023.
https://doi.org/10.1007/978-3-031-51300-8_3

economic, social and cultural relations between individuals of different nationalities. Such an issue led to diversifying the skills and expertise necessary to perform business and acquire cultures and thus achieve the required objectives (Al-Abrrow et al. 2021). In recent years, the use of information systems has increased in most matters of life, some of them used them in daily matters and others used them in organizations through the collection and processing of information, as the era of the information revolution depends directly on linking advanced information systems and how to manage their uses (Aboulola 2021; Albahri et al. 2021). Management information systems a major role in organizing, classifying, storing and retrieving information in a timely manner, which helps decision makers to make appropriate, efficient and effective decisions, especially those related to human resources (Abadi and Al-Ardhi 2012; AL-Fatlawey et al. 2021). These advances in information systems have enabled working individuals to digitally record keeping, making it easier for employees to work in companies (Boateng 2007; Alnoor et al. 2022). The importance of international HRM lies in how it manages its multicultural human resources in global companies.

As a result of the continuous developments in the world today and as a result of the increasing innovation and social interaction in wide scape, modern concepts have emerged to expand the scope of human without the restrictions of time and space through which employees all over the world will be able to communicate and create through metaverse (Limano 2023; Alsalem et al. 2022). Metaverse increases the complexity of these social interactions and introduces a new value system George et al. (2021). Metaverse applications are virtual reality technologies that mimic real reality. The metaverse first appeared in Neal Stephenson's novel. Metaverse has been described as 3D applications that present a virtual environment similar to the real one. Many business models have focused on virtual reality applications, and these technologies have become an urgent need for many companies in the business world (Lee et al. 2021; Atiyah 2020). Culture also includes the expectations of the organization, the philosophy of experience, and the values that bind the organization together and are expressed in its self-image, internal actions, and interactions with the outside world. Organizational culture has proven to be an important driver of organizational effectiveness by creating an internal environment conducive to individual productivity (Umans et al. 2016; Atiyah 2023). This study aims to investigate the impact of information management systems on international human resources management. The current study also wants to reveal the strength of the relationship between the information system and human resources management by shedding light on the metaverse culture in the oil sector. In addition, we focus on the damage of the knowledge gap using the culture as a vacuum change (Klasmeier and Rowold, 2020; Fadhil et al. 2021). This is to recognize that the oil sector faces problems in the management of international human resources. This study uses a sample of individuals working in the international companies which are BP (a British company), Baker Hughes (an American company), and ENI (an Italian company) in southern Iraq.

2 Literature Review

According to Habeeb et al. (2023) information technology consists of hardware and equipment, software, communications and users. While human resources management practices are represented in human resources planning, recruitment, training and development, compensation and incentives, performance evaluation. The conclusions of that study confirmed the existence of interrelationships and influence between information technology and human resources management practices at the macro and micro levels. Al Harthi (2021) aims to study the impact of management information systems and human resource management practices. Hence, previous studies explained how management information systems can help retain talented employees. Moreover, there was a relationship between qualification and performance management through management information systems and also a positive relationship between qualification and retention of talented employees (Al-Hchaimi et al. 2022; Chew et al. 2023; Muhsen et al. 2023). The main challenges of human resource management are the lack of the necessary skills to carry out human resource management practices. Mshana (2022) argued that human resources managers not only consider the use of information systems as a support for strategic human resources tasks, but also as a technology that facilitates managers to do many administrative and technical tasks in organizations and will enable them to obtain a lot of knowledge. According to Ferric Limano (2023) digital culture provides many new opportunities in the future, especially in virtual work and digital networks. These metaverse features lead to automation and remote work as well as wireless networks and management algorithms, and a new job certainly appears in the metaverse world that requires new skills in society. Alyaa Omar Kamel Faraj (2021) investigated the effect of reality of digital culture practices among students and the study suggests firms should develop a strategic plan to promote digital culture.

3 Hypotheses Development

The emergence of the concept of management information systems for the first time was in the mid-fifties to the mid-sixties, when it appeared with the use of computers, as applications were primarily just processing records and numbers of documents and reports (Gatea and Marina 2016). The management information system is an abbreviation for three words (management, information, systems). In order to fully understand the term management information system, each term of these three words must be revealed. Management is as "a set of functions including (planning, decision-making, organizing, directing, and controlling) directed towards the organization's resources (human, physical, financial, information resources) in order to achieve organizational goals in an effective manner". Information is a set of data that has been administered and has merit for firms and individuals (Shakhorska and Medykorsky 2018). It can also be defined as "the data that is processed and presenting it in a form that helps in decision-making".

The system has been referred to as a set of interrelated elements that influence each other to achieve a specific purpose (Ackoff 1971). Based on systems theory, the system consists of several elements and sub-systems, and these systems have features and interact information with each other (Weissenberger et al. 2019; Hamid et al. 2021).

The management of international human resources is a complex process because of the increase of multinational companies, separation of firms in different countries, and have different nationalities (Dmour et al. 2017; Khaw et al. 2022). Griffin and Pustay (2008) defined management of international human resources as a group of activities acquired through continuous development in the selection of manpower to achieve the objectives of the international company. It is a process of training, attracting, and developing international human resources to achieve global goals for organizations. In recent years, management information systems have helped the organization to accomplish human resource work in many multinational companies (Boiko et al. 2019). The widespread use of information systems has had a significant impact on the way human resources are managed at the present time. The use of information systems has become common in the field of business. Management information systems are represented in the effective use of hardware and software that link human resource management to each other. Such programs helped to carry out human resource activities (planning, recruitment, training, development) and predicting long-term performance Normalini et al. (2012). Carrying out the most important human resource activities such as planning and forecasting of human resources efficiently through human resource systems (Manhal et al. 2023). Information systems help improve human resource applications based on information technology (Abbas et al. 2023; Bozanic et al. 2023). There is a positive relationship between information systems and human resource management, (Marlene and Carlos 2018; John and Jackson 2007; Haider and Sundin, 2019; Swanson 2017; Gregor and Hevner 2013; Rainer et al. 2020). Previous studies revealed there is impact of unstable environments on international human resource management (Dickmann et al. 2017; Ramirez et al. 2016). There is a knowledge gap in human resource management in the economic context related to the role of talent management. Ongoing research should understand the role of international human resources management in talent management of different nationalities or different countries in terms of industry 5.0 such as adopting virtual reality applications. On the basis of the previous discussion, the following hypotheses were proposed:

H1: There is a positive influence between management information systems and International human resource management.

The interest in culture increased rapidly in the 1980 due to the competition of American companies that compete with Japanese companies in the field of electronics. Culture affects staff' behavior. Information system make managers and other employees communicate inside and outside the company along with knowledge of how to carry out their work (Lapsley and Rekers 2017; Cadez Guilding 2008; Tillmann and Goddard 2008; Emsley 2005). In addition, managers and employees will better understand information needs within organizations. Shared norms and values lead to an innovation-oriented culture which can be explained as the pursuit and experimentation of innovative ideas, the search for new business opportunities and the acceptance of higher levels of risk (O'Reilly 1991). These organizations are more likely to accept new ideas in management and non-management systems practices (Baird et al. 2018; Gupta and Salter 2018). Research in the Information system literature on the role of culture has been very limited, for example (Baird et al. 2018; Ax and Greve 2017; Zhang et al. 2015). Previous studies supported the role of culture in implementing information systems practices through the direct influence of networks (Hall 2010; Luft 2009; Davila and Wouters 2007; Van der Veeken and Wouters 2002; Bruns and McKinnon 1993; Yigitbasioglu 2016; Atiyah and Zaidan 2022). Managers in results-driven companies are primarily driven by achievement, actions, and high-performance expectations. In the past years, the selection of international human resources depended on the selection of managers to adapt across cultures (Pucik 1984). Effective international human resources managers need to share core metaverse cultural values and learn interdisciplinary and inter functional problem-solving. (Lapsley and Rekers 2017; Cadez Guilding 2008; Tillmann and Goddard 2008; Emsley 2005). Competitive culture theory is one of the components of the new international dimension of multicultural human resource management (John 2016). Cultural and institutional forces interact with international human resources (Sparrow and Makram 2015). Also, previous studies focused on the prevalence of human resources incentive policies and practices in Chinese multinational companies in the United States. Previous research also explored the impact of culture on human resource management practices in China and other countries around the world (Cooke et al. 2012). There is influence of cultural values on western human resource practices in African organizations (Warner 2010). There is a positive effect between culture and international human resource management. Based on the previous discussion, the following hypotheses were proposed:

H2: The relationship between management information systems and International human resource management will be strong when metaverse culture is high.

4 Methodology

The data were obtained from three international oil companies affiliated to the Basra Oil Company in Basra Governorate southern Iraq. This is due to the fact that the three companies contain international human resources of different nationalities. The sample size was 200 managers and officials out of 422 respondents. Six demographic variables were used (Gender, marital status, age, educational qualification, years of service, nationality). Three variables were used in this study as shown in Fig. 1.

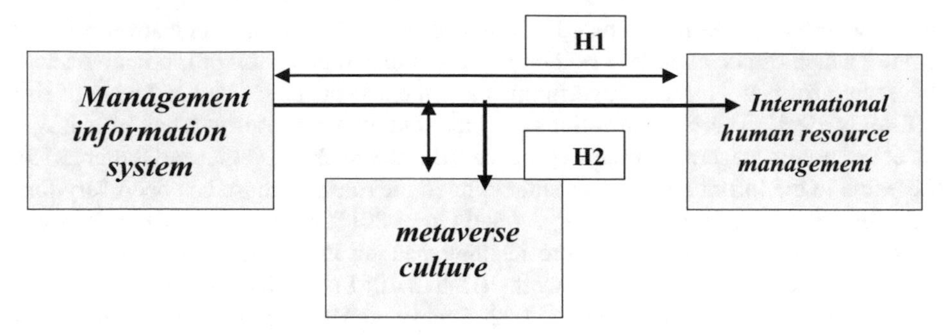

Fig. 1. Conceptual framework.

Management information systems scale was adopted (O' Brien and Marakas 2011). Which consists of 5 sub-dimensions. For example, the phrase "the company provides computers to employees and managers of all nationalities" was used to measure MR. Secondly, we used measure human resource, "employees easily communicate with each other". Third, we used measure DB, "the company provides the necessary software for work". Fourth, we used measure CN, "the available networks are commensurate with business needs". Fifth, we used measure SO. "The company's data system is subject to continuous updating and modification". Metaverse culture (MC), The Reality of Promoting Digital Culture [PDC]. Obstacles to Implementing Digital Culture [IDC]. Efforts To Develop a Digital Culture [DDC]. Which consists of 3 sub-dimensions. For example, the phrase "Promoting Digital Culture act as a motivating factor for mastery of work" was used to measure PDC. Secondly, was used to measure IDC, "There is a conviction among employees of the importance of participating in decision-making processes". Thirdly, was used to measure DDC, "The prevailing norms among employees helped to adapt among them". International human resource management (IHRM), international human resource planning [IHRP], international human resource recruitment and training [IHRRT], incentives and rewards for international human resources [IHRIR]: A scale was adopted (Feng 2016; Omar 2017). Which consists of 3 sub-dimensions. For example, the phrase "The company has plans to provide international human resources" was used to measure IHRP. Secondly, was used to measure IHRRT. "Have fair training and employment policies". Third, was used to measure IHRIR. "There are differences in the salary and bonus system in the company and other companies".

5 Data Analysis

One of the most important determinants of using appropriate statistical tools is the nature of the data distribution. Parametric statistical tools will be most appropriate for normally distributed data, while non-parametric statistical tools will be most appropriate for abnormally distributed data (Field 2009). We will do this by testing skewness and kurtosis to check the nature of the data distribution. Based on this test, and at the level

Table 1. Results of data distribution.

	Skewness			Kurtosis		
	Statistic	Std. Error	Z Skewness	Statistic	Std. Error	Z Kurtosis
MIS	−0.284	0.172	−1.651	−0.025	0.342	−0.072
IHRM	0.096	0.172	0.558	−0.187	0.342	−0.547
MC	0.260	0.172	1.512	−0.191	0.342	−0.557

Note: MIS = Management information system; IHRM = International human resource management; MC = Metaverse culture

of significance in this study is (0.05), the distribution is normal when the Z value for wobble and flattening does not exceed (±1.96). The Z value is extracted by dividing the calculated Skewness and Kurtosis values by their standard error (Kerr et al. 2002). Table 1 shows the results of this test.

Depending on the confirmatory factor analysis (CFA), validity and reliability were verified. Validity will be verified by ascertaining indicators of convergent validity, which determines the extent to which the sub-concepts (dimensions) that measure a concept are close to each other. The indicators of convergence validity are: (1) The standard saturation values (Factor Loading-FL) for each question of the scale, which must exceed (0.50), and it is better to exceed (0.70). (2) The average value of the extracted variance (AVE), which should be greater than (0.50) (Hair et al. 2010). On the other hand, reliability refers to the probability or extent to which the same results appear when negative measures are used at another time. The stability will be checked by verifying the two values of Composite Reliability and Cronbach's α stability coefficient, which can be obtained through the outputs of the confirmatory factor analysis. The stability of the three measures will be achieved by exceeding the value of the composite stability and Cronbach's alpha of (0.70). Table 2 shows the results reached.

Depending on the results shown in the above table, the questions of the dimensions of the independent variable (management information systems) have exceeded their factors loading (0.50). Also, the average variance extracted (AVE) for those dimensions exceeded (0.50), and this gives an indication of the validity of the two indicators of convergence for this variable. In addition, the two reliability values exceeded (0.70), which achieves reliability for the variable. Also, the questions of the dimensions of the dependent variable (international human resources management) exceeded their factors loading (0.50), except for two questions related to the compensation and rewards of international human resources management dimension (ihrt7 and ihrt8). Also, the (AVE) for those dimensions exceeded (0.50), and this gives an indication of the validity of the two indicators of convergence for this variable. In addition, the two reliability values exceeded (0.70), which achieves reliability for this variable. Finally, the questions of the dimensions of the moderator variable (metaverse culture) exceeded their factors loading (0.50). Also, the (AVE) for those dimensions exceeded (0.50), and this gives an indication

Table 2. Validity and reliability.

Variables	Factors	Items	FL	AVE	CR	Cronbach's α
MIS	HR	hr1	0.766	0.645	0.779	0.825
		hr2	0.763			
		hr3	0.876			
	MR	mr1	0.789	0.585	0.712	0.798
		mr2	0.792			
		mr3	0.71			
	SO	so1	0.733	0.59	0.718	0.708
		so2	0.736			
		so3	0.831			
	CN	cn1	0.865	0.577	0.703	0.815
		cn2	0.703			
		cn3	0.7			
	DB	db1	0.827	0.616	0.748	0.755
		db2	0.797			
		db3	0.728			
IHRM	IHRP	ihrp1	0.707	0.646	0.825	0.927
		ihrp2	0.843			
		ihrp3	0.839			
		ihrp4	0.819			
	IHRRT	ihrt1	0.668	0.622	0.86	0.882
		ihrt2	0.858			
		ihrt3	0.793			
		ihrt4	0.849			
		ihrt5	0.68			
		ihrt6	0.86			
		~~ihrt7~~	~~0.249~~			
		~~ihrt8~~	~~0.378~~			
	IHRIR	ihrc1	0.758	0.599	0.782	0.901
		ihrc2	0.803			
		ihrc3	0.763			
		ihrc4	0.771			
MC	PDC	ov1	0.738	0.271	0.528	0.885
		ov2	0.519			
		ov3	0.736			
	IDC	ob1	0.616	0.28	0.552	0.902
		ob2	0.725			
		ob3	0.705			
	DDC	on1	0.732	0.312	0.675	0.892

Note: MIS = Management information system; IHRM = International human resource management; MC = Metaverse culture

of the validity of the two indicators of convergence for this variable. In addition, the two reliability values exceeded (0.70), which achieves reliability for the variable.

Table 3 presents the results of the descriptive statistics and the correlation coefficient between the variables. The results indicate that most of the arithmetic mean ranges from agreement to some extent to agreement and beyond the hypothetical mean (3). On the other hand, the results indicate that the standard deviations were relatively few, which indicates a small dispersion in the data, and this supports the accuracy of the results. Finally, most of the correlation coefficients between the three variables and their dimensions were positive and statistically acceptable.

Table 3. Descriptive statistics and correlation coefficient

	HR	MR	SO	CN	DB	IHRP	IHRT	IHRC	MIS	IHRM	MC
HR	1										
MR	.347**	1									
SO	.309**	.456**	1								
CN	.442**	.286**	.340**	1							
DB	.368**	.269**	.419**	.485**	1						
IHRP	.170*	0.04	0.133	.356**	.306**	1					
IHRT	.160*	.241**	.176*	.343**	.239**	.503**	1				
IHRC	0.072	0.047	0.067	.264**	0.053	.362**	.402**	1			
MIS	.711**	.648**	.723**	.724**	.721**	.287**	.325**	.142*	1		
IHRM	.166*	0.12	.152*	.405**	.246**	.804**	.754**	.785**	.308**	1	
MC	.463**	.319**	.344**	.486**	.430**	.278**	.211**	.243**	.580**	.316**	1

Note: MIS = Management information system; IHRM = International human resource management; MC = metaverse culture

The current study model includes three variables, management information systems as an independent variable, international human resources management as a dependent variable, and organizational culture as a moderator variable. On this basis, the hypotheses will be tested through multiple regression analysis in the SPSS program. According to this analysis, the hypothesis is accepted or rejected based on the base values of the t and p values. To accept any hypothesis, the Critical ratio-t value must be greater than 1.96. While the p-value, which represents the level of reliability or acceptable error, must be less than 0.05. Table 4 shows the results of testing the two main hypotheses, while Table 5 shows the results of testing the sub-hypotheses.

The hypothesis that indicates a positive impact relationship of management information systems in international human resource management was accepted (H1: $p < 0.05$). The second main hypothesis is also accepted, which indicates the existence of a moderator role of metaverse culture in enhancing the relationship of the influence of management information systems in international human resource management (H2: $p < 0.05$). Finally, the coefficient of determination for the second main hypothesis was

Table 4. Testing the main hypotheses.

Model		Unstandardized Coefficients		Standardized Coefficients	t	Sig
		B	Std. Error	Beta		
H1	(Constant)	2.332	0.246		9.468	0.000
	MIS	0.320	0.070	0.308	4.558	0.000
H2	(Constant)	5.182	0.917		5.649	0.000
	MIS	−0.641	0.267	−0.617	−2.401	0.017
	MC	−0.893	0.329	−1.085	−2.714	0.007
	INT	0.298	0.090	1.890	3.297	0.001

a. Dependent Variable: IHRM

Note: MIS = Management information system; IHRM = International human resource management; MC = Metaverse culture

Table 5. Testing the sub hypotheses.

Model		Unstandardized Coefficients		Standardized Coefficients	t	Sig
		B	Std. Error	Beta		
(Constant)		2.344	0.247		9.478	0.000
H1a	HR	0.121	0.052	0.131	2.305	0.010
H1b	MR	0.002	0.063	0.002	0.026	0.980
H1c	SO	0.000	0.054	0.001	0.008	0.994
H1d	CN	0.285	0.059	0.284	4.808	0.000
H1e	DB	0.152	0.059	0.170	2.562	0.004
(Constant)		4.374	0.965		4.534	0.000
H2a	INT1	0.187	0.071	0.222	2.628	0.002
H2b	INT2	-0.120	0.092	-0.787	-1.300	0.195
H2c	INT3	0.039	0.082	0.279	0.475	0.636
H2d	INT4	0.249	0.078	0.356	3.193	0.001
H2e	INT5	0.181	0.076	0.225	2.391	0.018

a. Dependent Variable: IHRM

Note: MIS = Management information system; IHRM = International human resource management; MC = Metaverse culture

higher than the first main hypothesis, as it explained 16.9% of the changes in international human resource management.

Results In the Table 5, we can see that only three sub-hypotheses are accepted, related to the first main hypothesis, indicating the existence of a positive impact relationship between the components of management information systems related to human resources, communication networks, and databases in international human resource management. As for the sub-hypotheses of the second main hypothesis, three sub-hypotheses were also accepted, which indicate the existence of a moderator role of Metaverse culture in enhancing the impact relationship of the components of management information systems (human resources, communication networks, and databases) in international human resource management. Finally, the coefficient of determination for the sub-hypotheses of the second main hypothesis was higher than the hypotheses of the first main hypothesis, as it explained 25.1% of the changes in international human resources management.

6 Conclusion

The results showed that there is an impact and correlation between management information systems and international human resources management and the interactive role of Metaverse culture. In order to develop and expand large international companies, employees must develop their own ideas by adapting to technology in the workplace because it will facilitate the conduct of business and thus achieve superior profits through the use of digital networks at work and communication between managers in the home country and the host country. Increasing the awareness of officials in oil licensing companies about the metaverse culture, because problems and crises can occur to them if they cannot understand the different cultures of individuals within the company, and thus can reflect negatively on the company as a whole. In light of these recommendations, there is a set of implementation mechanisms that must be carried out by the Training and Development Department with the support of senior management through holding specialized and continuous workshops and training programs, focusing on reducing excellence in the organizational culture of international human resources in the companies concerned.

This study adopted the variable (Management Information Systems) as an independent variable, (International Human Resources Management) as a dependent variable, and (metaverse Culture) as an interactive variable, and worked on diagnosing these variables and studying them in oil licensing companies. About the cost, and in every study, there are always determinants exposed to it, and the determinants of this study were represented by the following points. Geographical determinants: represented by the locations of the oil licensing companies, where (BP) was in Rumaila, and (Baker Hughes) was in Rumaila as well, and (ENI) was in Zubair. Which required the researcher to make a great effort in order to go to more than one site in order to obtain information from the international human resources working in these international companies. The limited scientific sources of international human resource management, which made the researcher face difficulties in obtaining the largest amount of information that enriches the theoretical aspect of the current study. In this study, the impact of information systems on international human resources was measured. Future studies should measure the negative role of using Management information systems in international human resource management practices. This study was also applied in international companies. In future studies, it is preferable to apply it in local companies.

Appendix I

Management Information System human resources	
1	The company has a sufficient number of multiple employeeNationalities with competence in the field of management information systems
2	The company has a future vision in the field of information systems administrative
3	The company has a culture of management information systems
Material resources	
4	The company has modern tools, means and information technologies
5	The company is interested in maintaining technological devices and tools continuously
6	The company provides modern computers for all its employees
software	
7	The company relies on a variety of technological systems to perform its business
8	The company is interested in training its employees on computer software cutting edge
9	The company relies on the exchange of information through its advanced software
Networks	
10	The company provides a network (Wi-Fi) to facilitate Communication between employees at different levels
11	The company is interested in training its employees on computer software advanced and modern
12	The company relies on the exchange of information by software own modern
Databases	
13	The company has various databases
14	The company is constantly updating the databases
15	The employees can easily access the databases and quickly
International human resource planning	
1	Management provides an effective information system that serves the planning process Multinational HR
2	The company has future plans to provide human resources of their various nationalities
3	Management pre-determines a number of required skills that serve the business requirements of the company
4	Management determines its international human resource needsMulticulturalism capable of achieving competitive advantage
Recruitment and training international human resources	

(continued)

(*continued*)

Management Information System human resources	
5	The policies of attracting individuals in the company are fair and transparent
6	Competencies are recruited in the affiliated branches according to a policy The company's headquarters
7	Recruitment processes are subject to local laws and regulations
8	Flexible recruitment policies to gain new employees
9	Subsidiaries are influenced by the culture of the country in which they operate
10	There is a difference in employment policies compared to companies other international
11	Training and development programs are in line with resource qualifications human resources available in the company
12	The company is interested in training and development programs that increase experiences individuals to work
Compensation and rewards for international human resources	
13	Promotion and upgrading procedures in the company are fair and transparent
14	The salary and wages system in the company is developed and transparent
15	There are various incentives and rewards according to the work he performs individuals in the company
16	The company gives other wages for additional work
Metaverse culture The Reality of Promoting Digital Culture	
1	The company management can search the databases
2	Company management can use social media effectively
3	The ability to save and retrieve electronic information sources easily
Obstacles To Implementing Digital Culture	
4	Employees can use the phone or computer to help them get their work done
5	Employees can view the best training courses and seminars through computers and mobile phones
6	Google drive can be used to store and retrieve files
Efforts To Develop A Digital Culture	
7	The company's firewall is used to protect files and personal information
8	The company gives a list of regulations and procedures to employees showing them how to use the programs and applications
9	Information sources can be obtained through the company's digital stores

References

Abbas, S., et al.: Antecedents of trustworthiness of social commerce platforms: a case of rural communities using multi group SEM & MCDM methods. Electron. Commer. Res. Appl. **62**, 101322 (2023)

Ackoff, R.L.: Towards a system of systems concepts. Manage. Sci. .**17**(11) 661–671 (1971). P: 2

Al-Abadi, H.F., Al-Ardhi, J.K.: Information management systems a strategic perspective, Dar Safaa for publication and distribution, 1st edn Amman (2012)

Al-Abrrow, H., Fayez, A.S., Abdullah, H., Khaw, K.W., Alnoor, A., Rexhepi, G.: Effect of open-mindedness and humble behavior on innovation: mediator role of learning. International J. Emerg. Mark. **18**(9), 3065–3084 (2021)

Albahri, A.S., et al.: Based on the multi-assessment model: towards a new context of combining the artificial neural network and structural equation modelling: a review. Chaos Solitons Fractals **153**, 111445 (2021)

Al-Dmour, R.H., Masa'deh, R.E., Obeidat, B.Y.: Factors influencing the adoption and implementation of HRIS applications: are they similar? Int. J. Bus. Innov. Res. **14**(2), 139 (2017)

AL-Fatlawey, M.H., Brias, A.K., Atiyah, A.G.: The role of strategic behavior in achievement the organizational excellence" Analytical research of the manager's views of Ur state company at Thi-Qar governorate". J. Adm. Econ. **10**(37), 48–68 (2021)

Al-Hchaimi, A.A.J., Sulaiman, N.B., Mustafa, M.A.B., Mohtar, M.N.B., Hassan, S.L.B.M., Muhsen, Y.R.: Evaluation approach for efficient countermeasure techniques against denial-of-service attack on MPSoC-based IoT using multi-criteria decision-making. IEEE Access **11**, 89–106 (2022)

Alnoor, A., et al.: How positive and negative electronic word of mouth (eWOM) affects customers' intention to use social commerce? A dual-stage multi group-SEM and ANN analysis. Int. J. Hum. Comput. Interact. **4** (2022). https://doi.org/10.1080/10447318.2022.2125610

Alsalem, M.A., et al.: Rise of multiattribute decision-making in combating COVID-19: a systematic review of the state-of-the-art literature. Int. J. Intell. Syst. **37**(6), 3514–3624 (2022)

Atiyah, A.G.: The effect of the dimensions of strategic change on organizational performance level. PalArch's J. Archaeol. Egypt/Egyptol. **17**(8), 1269–1282 (2020)

Atiyah, A.G., Zaidan, R.A.: Barriers to using social commerce. In: Alnoor, A., Wah, K.K., Hassan, A. (eds.) Artificial Neural Networks and Structural Equation Modeling. Springer, Singapore (2022). https://doi.org/10.1007/978-981-19-6509-8_7

Ax, C., Greve, J.: Adoption of management accounting innovations: organizational culture compatibility and perceived outcomes. Manag. Account. Res. **34**, 59–74 (2017)

Baird, K., Su, S., Tung, A.: Organizational culture and environmental activity management. Bus. Strategy Environ. **27**(3), 403–414 (2018)

Boateng, A.: The Role of Human Resource Information Systems (HRIS) in Strategic Human Resource Management (SHRM) (thesis). Sweden: Swedish School of Economics and Business Administration, 112 p. (2007)

Boiko, A., Shendryk, V., Boiko, O.: Information systems for supply chain management: uncertainties, risks and cyber security. Procedia Comput. Sci. **149**, 65–70 (2019)

Bornay -Barrachina, M.: International Human Resource Management: How should employees be managed in an international context (2019). https://doi.org/10.4018/978-1-5225-5781-4

Bozanic, D., Tešić, D., Puška, A., Štilić, A., Muhsen, Y.R.: Ranking challenges, risks and threats using fuzzy inference system. Decis. Making Appl. Manage. Eng. **6**(2), 933–947 (2023)

Cadez, S., Guilding, C.: An exploratory investigation of an integrated contingency model of strategic management accounting. Account. Organ. Soc. **33**(7–8), 836–863 (2008)

Chew, X., Khaw, K.W., Alnoor, A., Ferasso, M., Al Halbusi, H., Muhsen, Y.R.: Circular economy of medical waste: novel intelligent medical waste management framework based on extension linear Diophantine fuzzy FDOSM and neural network approach. Environ. Sci. Pollut. Res. **30**, 1–27 (2023). https://doi.org/10.1007/s11356-023-26677-z

Cooke, F.L.: The globalization of Chinese telecom corporations: strategy, challenges and HR implications for the MNCs and host countries. Int. Hum. Resour. Manage. **23**(9), 1832–1852 (2012)

Dickmann, M., Parry, E., Keshavjee, N.: Localization of staff in a hostile context: An exploratory investigation in Afghanistan. Int. J. Hum. Resour. Manage. **5192**, 1–29 (2017). https://doi.org/10.1080/09585192.2017.129153

Emsley, D.: Restructuring the management accounting function: a note on the effect of role involvement on innovativeness. Manag. Account. Res. **16**(2), 157–177 (2005)

Fadhil, S.S., Ismail, R., Alnoor, A.: The influence of soft skills on employability: a case study on technology industry sector in Malaysia. Interdiscip. J. Inf. Knowl. Manag. **16**, 255 (2021)

Feng, L.: An assessment of international human resource management (IHRM) Practices in Chinese Multinational Corporations (MNCs) in Africa: Standardization or Adaptation vol. 42 (2016)

Field, A.: Discovering Statistics Using SPSS, 3rd edn (2009)

Gatea, A.A., Marina, V.: Higher education funding in Iraq in terms of the experience of particular developed countries. Int. J. Adv. Stud. **6**(1), 8–17 (2016)

George, A.H., Fernando, M., George, A.S., Baskar, T., Pandey, D.: Metaverse: the next stage of human culture and the internet. Int. J. Adv. Res. Trends Eng. Technol. (IJARTET) **8**(12), 1–10 (2021). ISSN 2394–3777 (Print) ISSN 2394–3785

Griffin, M., Pustay. Business Research Methods 8th (Eight) Edition. South-Western College Pub, New Castle (2008)

Gupta, G., Salter, S.B.: The balanced scorecard beyond adoption. J. Int. Account. Res. **17**(3), 115–134 (2018)

Hang Lee, L., et al.: All one needs to know about Metaverse: a complete survey on technological singularity, virtual ecosystem, and research agenda. J. LaTEX IEEE **11**(8), 1–66 (2021)

Haider, J., Sundin, O.: Invisible search and online search engines: The ubiquity of search in everyday life. Routledge (2019)

Hair, J.F., Black, W.C., Babin, B.J., Anderson, R.E.: Multivariate Data Analysis. 7th ed. Pearson Prentice Hall (2010)

Hall, M.: Accounting information and managerial work. Account. Organ. Soc. **35**(3), 301–315 (2010)

Hamid, R.A., et al.: How smart is e-tourism? A systematic review of smart tourism recommendation system applying data management. Comput. Sci. Rev. **39**, 100337 (2021)

John, N.N.: Cultural dimensions in global human resource management: implications for Nigeria. Ugoani college of management and social sciences, Rhema University, Nigeria 2016

John, N.N.: Ugoani College of management and social sciences, Rhema University, Nigeria, independent journal of management & production (IJM&P). http://www.ijmp.jor.brv. 7, n. 3, July–September 2016 ISSN: 2236–269X. https://doi.org/10.14807/ijmp.v7i3.429

Kamel Faraj, A.O.: Developing the digital culture among the students of educational faculties in prince sattam bin abdulaziz university. Int. J. High. Educ. **10**(3), 158–158 (2021)
Kerr, A.W., Hall, H.K., Kozub, A.K.: Doing Statistics with SPSS. SAGE Publications, London (2002)
Khaw, K.W., et al.: Modelling and evaluating trust in mobile commerce: a hybrid three stage Fuzzy Delphi, structural equation modeling, and neural network approach. Int. J. Hum.-Comput. Interact. **38**(16), 1529–1545 (2022)
Klasmeier, K.N., Rowold, J.: A multilevel investigation of predictors and outcomes of shared leadership. J. Organ. Behav. **41**(9), 915–930 (2020)
Limano, E.: New digital culture metaverse preparation digital society for virtual ecosystem. In: E3S Web of Conferences, vol. 388, p. 04057 (2023). https://doi.org/10.1051/e3sconf/202338804057 ICOBAR 2022
Lapsley, I., Rekers, J.V.: The relevance of strategic management accounting to popular culture: the world of West end musicals. Manag. Account. Res. **35**, 47–55 (2017)
Luft, J.: Nonfinancial information and accounting: a reconsideration of benefits and challenges. Account. Horiz. **23**(3), 307–325 (2009)
Manhal, M., Al-khalidi, A., Hamad, Z.: Strategic network: managerial myopia point of view. Manag. Sci. Lett. **13**(3), 211–218 (2023)
Milliman, J., Von Glinow, M., Nathan, M.: Organizational fez cycles Nd strategic international Umar resource management in multinational Mopanis: implications or congruence theory. J. Acad. Manage. Rev. **16**(4), 318–339 (1991)
Mshana, G., et al.: Same habitus in new field? How mobile phone communication reproduces masculinities and gender inequality in intimate relationships in Mwanza, Tanzania. J. Soc. Pers. Relat. **39**(11), 3351-3372 (2022)
Muhsen, Y.R., et al.: Enhancing NoC-based MPSoC performance: a predictive approach with ANN and guaranteed convergence Arit Hmetic optimization algorithm. IEEE Access **11**, 90143–90157 (2023)
Normalini, Ramayah, T., Kurnia, S.: Antecedents and outcomes of human resource information system (HRIS) use. Int. J. Prod. Perform. Manage. **61**(6), 603–623 (2012)
O'Reilly, C.A., Chatman, J., Caldwell, D.F.: People and organizational culture: a profile comparison approach to assessing person-organization fit. Acad. Manag. J. **34**(3), 487–516 (1991)
Omar, K.A.: The Impact Of Selected Human Resource Management Practices On Performance of Public Water Utilities In Tanzania. p. 250 (2017)
Rainer, R.K., Prince, B., Cegielski, C.G.: Introduction to Information Systems, 5th edn. Wiley, Singapore (2020)
Ramirez, J., Madero, S., Muniz, C.: The impact of narcoterrorism on HRM systems. Int. J. Hum. Resour. Manage. **27**(19), 2202–2232 (2016). P. 25
Swanson, E.B.: Information systems. In: McDonald, J.D., Levine-Clark, M. (eds.), Encyclopedia of library and information science (4th ed.). CRC Press (2017)
Tillmann, K., Goddard, A.: Strategic management accounting and sense-making in a multinational company. Manag. Account. Res. **19**(1), 80–102 (2008)
Umans et al.: 'Effect of organizational culture on company's'. (thesis). Sweden: Swedish School of Economics and Business Administration 79–86 p. (2016)
Van der Veeken, H.J.M., Wouters, M.J.F.: Using accounting information systems by operations managers in a project company. Manag. Account. Res. **13**(3), 345–370 (2002)
Voorberg, S., Eshuis, R., van Jaarsveld, W., van Houtum, G.J.: Decisions for information or information for decisions? Optimizing information gathering in decision-intensive processes. Decis. Support Syst. **151**, 113632 (1996)

Warner, M.: In search of Confucian HRM: theory and practice in greater China and beyond. Int. J. Hum. Resour. Manage. **21**(12), 2053–2078 (2010)

Weissenberger-Eibl, M.A., Almeida, A., Seus, F.: A systems thinking approach to corporate strategy development. Systems **7**(1), 16 (2019). https://doi.org/10.3390/systems7010016,p:2

Yigitbasioglu, O.: Firms' information system characteristics and management accounting adaptability. Int. J. Account. Inf. Manag. **24**(1), 20–37 (2016)

Zhang, Y.Z., Hoque, Z., Isa, C.R.: The effects of organizational culture and structure on the success of activity-based costing implementation. Adv. Manag. Account. **25**, 229–257 (2015)

Impact of Metaverse at Workplace: Opportunity and Challenges

Bushra Al Harthy[1(✉)], Aseela Al Harthi[2], Arash Arianpoor[3], and Ali Shakir Zaidan[4]

[1] University of Higher Technology and Applied Science, Muscat, Oman
bushra.nasser.alharthy@gmail.com
[2] Sultan Qaboos University, Muscat, Oman
aseela@squ.edu.om
[3] Department of Accounting, Imam Reza International University, Mashhad, Iran
[4] School of Management, Universiti Sains Malaysia, Gelugor, Penang, Malaysia

Abstract. Metaverse represent a phenomenon in the workplace and propose a potential implication for the future of work environments. This paper examines the impact of metaverse at the workplace in terms of achieving work-life balance, enhancing job satisfaction, and improving employee performance. The results identified that there is a positive relationship between Metaverse at workplace and work life balance, job satisfaction and employee performance. This study offers valuable insights into the potential implications of the metaverse in the context of the workplace and proposes avenues for further investigation.

Keywords: Metaverse · Employee Productivity · Job Satisfaction · Work-life Balance · Effective Communication

1 Introduction

The concept of the metaverse has been delineated and expounded upon from multiple scholarly viewpoints. Ondrejka (2005) is widely regarded as the first individual to provide a definition for the metaverse. Metaverse as an online environment that serves as a substitute for the physical world, enabling users to engage in social interactions, perform economic activities, and seek entertainment. Ondrejka further emphasised the usage of the real world as a metaphor within this virtual realm. Numerous researchers have undertaken investigations regarding the prospective implications of the metaverse in the context of higher education (Aldowah, Rehman, Ghazal, & Umar, 2017). The primary aim of their research is to ascertain the fundamental factors that lead to the successful implementation of the metaverse. Consequently, extensive studies have been conducted in various academic contexts, primarily focusing on the metaverse. The utilisation of the metaverse was employed by researchers in an educational setting, with a specific emphasis on the application of a problem-based methodology. This methodology enables both students and educators to present a problem and examine various potential resolutions in a simulated setting. The use of three-dimensional classrooms and avatars enables this process (Barry et al., 2009; Kanematsu et al., 2013). Furthermore, extensive

M. Al-Emran et al. (Eds.): IMDC-IST 2024, LNNS 876, pp. 54–68, 2023.
https://doi.org/10.1007/978-3-031-51300-8_4

research has been conducted on the effects of the metaverse on various sectors, including business (Falchuk, Loeb, & Neff, 2018), healthcare (Wu, Lin, & Bowman, 2022), and tourism (Griol Barres, Sanchis de Miguel, Molina López, & Callejas, 2019; Manhal et al., 2023).

The primary objective of this research is to augment the current corpus of scholarly work by including the notion of the metaverse within the framework of the professional environment. The fundamental theoretical contribution of this research is to underscore the impact of the metaverse on the workplace, with the aim of enhancing employee work-life balance, job satisfaction, and employee performance. According to Alfaisal, Hashim, and Azizan (2022), the concept of the "metaverse" refers to a virtual domain that enhances the physical environment and actuality. Hopkins and Bardoel (2023) believe that the utilisation of the metaverse affords employees the opportunity to efficiently fulfil their professional obligations from diverse geographic places, hence fostering a more favourable balance between work and personal life. Moreover, Shravanthi et al. (2015) proposed that the adoption of a conducive work-life balance for personnel could potentially result in a reduction in employee turnover. Kaaria and Mwaruta (2023) have established a positive correlation between the existence of a metaverse within the workplace and increased levels of employee job satisfaction. The metaverse's capacity to foster a more agreeable work environment and reduce work-related stress is the reason behind this phenomenon. The adoption of metaverse technology facilitates the participation of employees in virtual collaboration and communication, hence leading to improved employee performance (Cho & Lee, 2022; Gatea and Marina, 2016).

The current body of literature related to the metaverse has not thoroughly examined the concepts of work-life balance, job satisfaction, and employee performance, as indicated by several studies conducted in this domain. The main objective of this research is to propose a theoretical framework that investigates the influence of metaverses on the professional environment, specifically emphasising aspects of work-life balance, job satisfaction, and employee effectiveness. Hence, this study attempt is expected to provide new and valuable insights on the metaverse and its impact on employees in a professional environment. The primary objective of the research is to provide a thorough comprehension of the effects of the metaverse on work-life balance, job satisfaction, and employee performance inside the workplace. The study's results will offer organisations a more profound understanding of the influence of the metaverse on the workplace, particularly on its capacity to enhance work-life balance, job satisfaction, and employee effectiveness. The research presented here aims to offer valuable insights on the implications of the metaverse within the employment setting. Moreover, it is imperative to understand the implications of the metaverse for individuals at the regional level. Therefore, gaining a thorough understanding of the implications of the metaverse will allow firms to develop suitable strategies for its implementation, ultimately enhancing employee performance and satisfaction.

2 Literature Review

Koohang et al., (2023) provides a definition of the metaverse as a virtual environment that incorporates several reality technologies, such as mixed reality, 3D graphics, and virtual reality. This immersive digital space enables real-time experiences and interactions that are not feasible within the confines of the physical environment. The metaverse developed by Recker, Lukyanenko, Jabbari Sabegh, Samuel, & Castellanos (2021) constitute a network-based world that offers consumers an immersive online experience. The metaverse can be described as a vast and interconnected network of virtual worlds, characterised by real-time 3D environments. Within this network, users have the ability to engage in synchronous and persistent experiences alongside countless other users (Abbas et al., 2023; Bozanic et al., 2023). Moreover, the metaverse ensures the seamless transfer of data, encompassing aspects such as identification, entitlements, objects, communications, and payments (Ball, 2022). According to Dionisio, Iii, & Gilbert (2013), the word "virtual reality" can be defined as a computer-generated environment that extends beyond the physical world. This environment is fully immersive, three-dimensional, and digitally created. It encompasses the entirety of the shared online area, incorporating all dimensions of representation. According to Dionisio et al. (2013), the metaverse evolved from being a collection of separate virtual worlds to a vast network of interconnected virtual worlds.

Within the realm of the Metaverse, individuals have the opportunity to partake in various social endeavors, including but not limited to engaging in discourse pertaining to a particular matter, cooperating on a shared undertaking, participating in recreational activities, and acquiring knowledge through the process of experiential learning or problem-solving (Hwang & Chien, 2022). Indeed, there are notable distinctions between convening within the Metaverse digitally and engaging in virtual meetings through other video platforms such as Google Meet. When individuals utilise the metaverse as a platform for meetings, they have the ability to engage in a multitude of activities through their avatars, extending beyond the confines of mere meetings (Hatane, Sondak, Tarigan, Kwistianus, & Sany, 2023). In recent times, other corporations have also been endeavouring to gain access to the Metaverse. Owing to the substantial investments made by large corporations. For instance, certain accounting firms have established their physical presence within virtual reality environments, such as the first CPA company operating within Decentraland and PwC Hong Kong within the Sandbox (Maurer, 2022). Based on the above the metaverse can be seen as virtual network beyond the physical environment where devices and objects are connected and interacted with the internal and the external environment as demonstrated in Fig. 1.

Although there are many elements under the metaverses as depicts in Fig. 2, this paper aims to elucidate the impact of implementing the metaverse at the workplace.

The metaverse is conceptualised as a virtual reality scope wherein individuals engage in communication and interaction within a communal setting. Within the context of the workplace, the metaverse has the potential to bring about a transformative shift in the manner in which individuals engage in their daily tasks and foster collaboration. The incorporation of the metaverse in the workplace has a wide range of effect for both people and organisations. This section will elucidate the impact of the metaverse at the workplace in terms of work-life balance, job satisfaction and employee performance.

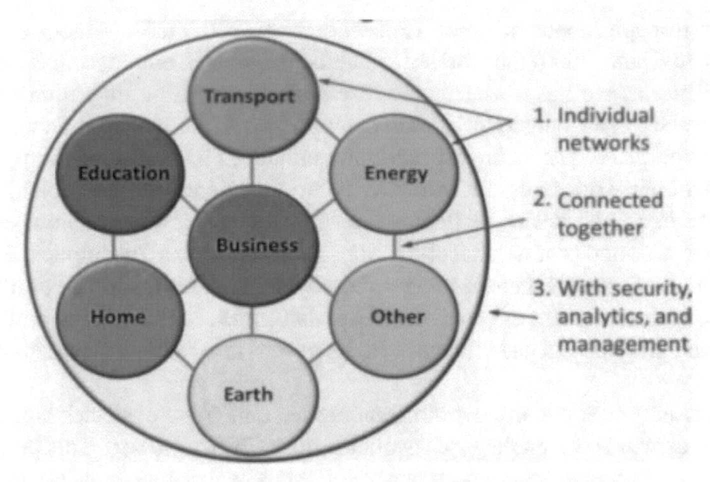

Fig. 1. Metaverse at Workplace. Source: Cisco IBSG, April (2011)

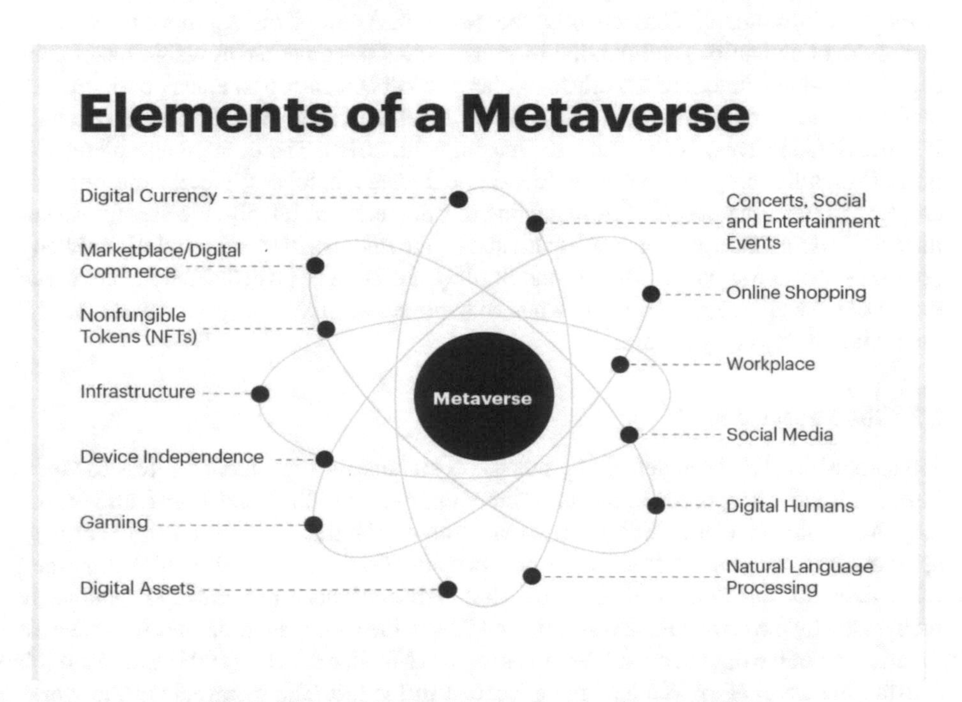

Fig. 2. Elements of a Metaverse. Source: (Gartner, 2023)

2.1 Work-Life Balance

A work-life balance is when employees have a proper balance between work obligations and personal commitments without any stress (Igbinomwanhia et al., 2012). The work-life balance refers to "the relationship between the institutional and cultural times and spaces of work and non-work in societies where income is predominantly generated and distributed through labour markets" (Felstead et al., 2002, p.56; Alnoor et al., 2022). According to Clark (2000) the work-life balance is when an employee has his/her work responsibilities as well as his/her home responsibilities with a minimum of role conflict. Metaverses contribute to the improvement of work-life balance among employees within the workplace. The metaverse facilitates enhanced flexibility, enabling employees to carry out their work remotely from any location. Therefore, the implementation of a metaverse in the workplace facilitates the achievement of a harmonious equilibrium between personal and professional obligations for employees. This implies that employees are able to effectively manage their personal and work obligations. For instance, the simultaneous management of caregiving responsibilities, such as daycare, while ensuring sustained involvement and efficiency in the professional setting (Hopkins & Bardoel, 2023).

In the twenty-first century, workers encounter numerous obstacles and experience heightened demands within the professional setting (Belwal, 2019). This pertains to the escalation in job obligations. Consequently, there has been a noticeable trend of employees dedicating more extensive periods of time to work-related activities beyond their domestic environment. Consequently, the comprehension of the equilibrium between professional obligations and familial responsibilities has emerged as a significant concern. The establishment of a work-life balance has been shown to enhance the overall quality of life for employees and contribute to the effectiveness of organisations (Leitão, Pereira, & Gonçalves, 2019; Peters & Heusinkveld, 2010). Therefore, the implementation of work/life programs has been found to enhance employees' morale and improve employee retention rates, while also leading to a decrease in staff absenteeism (Shravanthi et al., 2015). This assertion is substantiated by the findings of Mendis and Weerakkody (2017) as well as McDonald, Brown, and Bradley (2005), who posit that employees experience enhanced mental well-being when they possess a greater degree of autonomy in managing their work schedules.

2.2 Job Satisfaction

Job satisfaction has been defined by many researchers. Job satisfaction refers to "feelings or affective responses to facets of the situation" (Smith, Kendall & Hulin, 1969, p.6). According to Weiss (2002), job satisfaction can be described as a positive or negative evaluative opinion of one's job or work situation'. Organ (1988, p.4) described job satisfaction as "individual behaviour that is discretionary, not directly or explicitly recognised by the formal reward system and that in the aggregate promotes the effective functioning of the organization". Whereas, Agho, Mueller & Price (1993) explained job satisfaction as an employee having an interest in his job. The metaverse at the workplace can help to minimize the stress at work by providing a more comfortable work environment. This will lead to enhanced job satisfaction (Alharbi and Alnoor, 2022; Kaaria & Mwaruta, 2023). Employees who are satisfied with their job tend to stay in their work for a long period while dissatisfied employees tend to leave their jobs for other jobs or be absent (Kohler & Mathieu, 1993; Saari & Judge, 2004). Several studies have demonstrated that when an employees' satisfaction is high, their performance will also be high. In short, there is a positive relationship between job satisfaction and employee performance (Judge, Thoresen, Bono, & Patton, 2001). Moreover, Wilkinson

(1992) emphasised that there is a positive relationship between job satisfaction and the level of efficiency and quality of work (Muhsen et al., 2023; Al-Hchaimi et al., 2022; Chew et al., 2023).

Also, job satisfaction improves an employee's innovation and relaxation, enabling them to do the work. Hence, it reduces employee tension that caused a gap between his expectations of the job and unmet needs. As highlighted by Aziri (2011), job satisfaction results in a sense of enthusiasm and happiness in an employee's work. Usually, satisfied workers are seen as more productive, creative and committed to their workplace. Therefore, job satisfaction is considered as one of the important issues for organizations (Bhatti & Qureshi, 2007). In addition, Potkány & Giertl (2013) pointed out that job satisfaction plays a crucial role in an employee's working life, in the sense of motivation, performing their work efficiently, as well in their as mental health. Employee satisfaction is vital, not only for organizational success but also for helping to enhance an employee's health.

2.3 Employee Performance

Employee performance can be defended as the employee outcome at workplace based on his duties at workplace (Aliyyah et al., 2021; Alsalem et al., 2022). According to Rizwan, Tariq, Hassan, & Sultan (2014) employee performance is seen as what the employee does and does not do at the workplace. Aliyyah et al. (2021) highlighted employee performance as the result of work achieved by an employee on the basis of skill, experience, and time. The metaverse at the workplace helps enable employees to work together as if they were physically present in the same room via virtual meetings, team-building activities, and project collaborations. The virtual collaboration helps boost communication among the staff, strengthens team cohesion, and fosters creativity. Consequently, employee productivity and efficiency will be improved (Cho & Lee, 2022).

In addition, the metaverse has distinct prospects for immersive and captivating training programs. In contrast to conventional training approaches, which may have inherent limitations, the metaverse offers interactive and immersive training simulations that provide employees the opportunity to engage with genuine circumstances, thereby augmenting their skills and knowledge. This practice not only conserves time and resources, but also empowers employees to acquire and implement novel knowledge within a secure and regulated virtual setting (Yemenici, 2022).

3 Research Methodology

The technique employed in this research is predicated upon a comprehensive examination of existing scholarly literature related to the metaverse, work-life balance, job satisfaction, and employee performance. The primary emphasis of this paper centers on the metaverse's implications within the context of employment. Therefore, the focus is on three key aspects: achieving a balance between work and personal life, enhancing job satisfaction, and improving employee performance. The primary domain of inquiry that warrants significant attention in the field of metaverse research initial inquiry suggests that the incorporation of a supplementary database would result in minimal enhancement in the quantity of articles related to the metaverse. Following that, the search encompasses articles that incorporate the term "metaverse." Google Scholar, Emerald Insight, and Springer are widely recognised academic databases that offer extensive access to a diverse array of scholarly articles and research papers. Researchers, scholars, and students frequently utilise these platforms as a means of accessing scholarly literature that has been critically examined and deemed authoritative in many academic disciplines. Consequently, a search strategy involving the use of keywords was employed to locate relevant scholarly publications relating to the integration of the metaverse within the context of employment.

Subsequently, the outcomes were categorised based on the correlation between the metaverse and work-life balance, the metaverse and job satisfaction, and the metaverse and employee performance. The subsequent stage of the study involves an examination of the articles within the designated category, with a particular focus on the exclusion of publications without peer review and those not published in the English language. This technique ensures the quality of academic work. Themes that exhibit recurrent patterns across many databases are likewise excluded. To guarantee the retrieval of highly relevant articles, it is important to include the term "metaverse" at least once within the entirety of the selected articles. After conducting a thorough review of the literature, any journal articles that are deemed irrelevant are excluded, ensuring that all selected publications align with the precise focus of this research. The current research incorporates supplementary articles using both forward and backward citations, including journal articles that were not encompassed in the first database search.

4 Result

Based on a comprehensive analysis of the existing literature pertaining to this subject, it is evident that there exists a favourable correlation between managerial involvement in the workplace and various outcomes such as work-life balance, job satisfaction, and employee performance.

4.1 Work-Life Balance

According to the findings of Ellenbecker (2004), it was observed that employees exhibit a preference for flexibility in order to effectively manage the demands of both their personal and professional domains. According to George (2015), the establishment of this balance is expected to enhance employees' dedication within the organisational setting. Therefore, the establishment of a favourable balance between professional and personal domains within organisations has been found to enhance employee retention (Leners, Roehrs, & Piccone, 2006; AL-Fatlawey et al., 2021). According to Ellenbecker's (2004) research, contemporary employees exhibit a greater inclination to remain with an organisation for a prolonged duration when provided with flexible work schedules that enable them to effectively manage their personal and professional commitments. Furthermore, Loan-Clarke et al. (2010) found that when an organisation offers employees the opportunity to fulfil family responsibilities, there is an increase in employee retention.

The concept of work-life balance, as previously examined, enables individuals to achieve a balance between their devotion to the organisation and their personal responsibilities. Furthermore, research has indicated that the implementation of a work-life balance has a positive impact on employee engagement, job satisfaction, and overall job performance (Igbinomwanhia et al., 2012). Hence, fostering a strong relationship between employees and organisations can yield benefits such as the achievement of organisational goals and improved employee retention (Garg & Yajurvedi, 2018; Lazar, Osoian, & Ratiu, 2010). Based on a study conducted in the United Kingdom, it was shown that those who are able to maintain a positive work-life balance demonstrate a higher probability of retaining their jobs, effectively managing tasks, displaying increased loyalty, and exhibiting higher levels of productivity. The above-mentioned outcomes have advantageous implications for companies, as they contribute to the effective execution of strategies and policies (Igbinomwanhia et al., 2012; Ojo, Salau, & Falola, 2014; Sadaa et al., 2022). Additionally, it is imperative to acknowledge that organisations have the capacity to achieve economic benefits through the provision of a beneficial work-life balance. The previously mentioned observation holds great importance as numerous organisations face financial challenges due to the considerable costs incurred from both direct and indirect expenditures linked to high personnel turnover (Harhara, Singh, & Hussain, 2015). The preservation of an appropriate balance between work and personal life is frequently regarded as an essential component in the retention of employees.

The integration of metaverse technology in the workplace introduces a potential concern regarding the maintenance of work-life balance and the occurrence of burnout. This arises from the possibility of boundaries being indistinct between professional responsibilities and personal life commitments. Insufficient implementation of work-life balance strategies may result in employees encountering difficulties in disengaging from work, hence exacerbating stress levels and fostering burnout. It is imperative for organisations to actively endorse and facilitate activities that promote a harmonious equilibrium between work and personal life (Park, Ahn, & Lee, 2023; Atiyah, 2022). Based on the above discussion, it may be concluded that hypothesis one has been validated.

4.2 Job Satisfaction

Employee satisfaction is seen as a valuable factor in reducing employee turnover in organizations (Kibui, Gachunga & Namusonge, 2014). Also, the effort and their commitment of satisfied employees are vital to organizational success (Arif & Farooqi, 2014). Further, employees who have high job satisfaction tend to be more productive in reaching organizational goals and policies, as well as increasing their loyalty (Kim, Knutson, & Choi, 2016; Sari, 2018). In organizations, behavioural job satisfaction is considered as a crucial aspect (Pandey & Khare, 2012; Atiyah and Zaidan, 2022). Thus, most organizations try to ensure they create a satisfied workforce because satisfied employees will make more effort to reach organizational goals (Arif & Farooqi, 2014; Pushpakumari, 2008).

According to Bakotić (2016), employees that experience satisfaction in their work tend to exhibit higher levels of creativity, commitment, and performance, so contributing to the overall success of the organisation. Furthermore, numerous empirical investigations have been undertaken to explore the correlation between job satisfaction and organisational commitment. Gunlu, Aksarayli, and Perçin (2010) conducted a research study with the aim of examining the impact of job satisfaction on organisational commitment among managers in large-scale hotels located in the Aegean area of Turkey. The findings of the study indicate that job satisfaction exhibits a substantial influence on both normative and emotional commitment.

Also, employees with a high level of satisfaction are motivated to be more engaged in all working processes, having a high commitment toward their work, and this helps to improve employee retention (Deshpande, Arekar, Sharma, & Somaiya, 2012; Kibui et al., 2014; Patriota, 2009). Thus, based on Kaaria & Mwaruta (2023) found that metaverse at workplace help employees to feel high level of engagement and job satisfaction. Hence, metaverse has a positive effect at workplace as it helps to improve employee job satisfaction. Therefore, hypothesis two is supported in the present investigation.

4.3 Job Satisfaction

The assessment of an employee's performance includes both their active behaviours and their passive omissions. Employee performance encompasses various crucial elements, such as the quantity and quality of output, regularity and punctuality in attendance, flexibility and interpersonal skills, and promptness in completing assignments (Shahzadi et al., 2014). According to the research conducted by Yang (2008), the results suggest that the validation of employee performance is not feasible. According to Yang (2008), businesses possess the opportunity to adopt direct incentives and prizes that are linked to employee performance, especially in cases where employee performance can be readily observed. In a study conducted by Yang (2008), the author examined the impact of acknowledging, recognising, and rewarding employee success on varying levels of employee productivity. The findings of the study revealed the significant influence played by these factors in differentiating levels of employee productivity. The effectiveness of an organisation's performance and its reward management system exert a substantial influence on the morale and productivity of its personnel (Güngör, 2011; Al-Abrrow et al., 2019). Companies usually devote substantial resources to meeting the demands of their clients, yet they often overlook the significance of guaranteeing employee satisfaction. Nevertheless, it is well acknowledged that the level of pleasure experienced by consumers is dependent on the level of satisfaction among the staff members. Ahmad, Wasay, and Ullah (2012) claim that employee satisfaction has the potential to positively impact productivity levels, hence contributing to improved customer satisfaction. The influence of motivation on employee performance is a topic of considerable importance, as highlighted by Paais and Pattiruhu (2020). According to Shahzadi et al. (2014), increased motivation among employees is associated with higher levels of effort exerted in their respective job roles, which subsequently contributes to a notable enhancement in overall performance.

The metaverse enhances remote cooperation by offering rich and captivating virtual experiences. It serves as a means of connecting team members who are physically separated, facilitating efficient collaboration irrespective of their geographical dispersion. Metaverse at the workplace is augmented by the ability for employees to effortlessly transition between remote work and virtual collaboration within the metaverse, hence amplifying flexibility and efficiency (Hwang, Shim, & Lee, 2022). Traditional office spaces have been revolutionised by the introduction of virtual work environments, which provide the flexibility to be customised and personalised according to individual preferences. Employees have the ability to design and customise their work environments according to their individual interests and requirements (Boughzala, de Vreede, & Limayem, 2012). The virtual workplace paradigm fosters a heightened sense of ownership and empowerment, hence resulting in amplified productivity and enhanced job satisfaction (Hwang, Shim, & Lee, 2022).

The implementation of a metaverse at the workplace has the potential to enhance staff productivity. With the ability to adapt to various work situations that are most beneficial to their productivity (Boughzala, de Vreede, & Limayem, 2012). In terms of style, employees have the ability to mitigate distractions and enhance their task-oriented focus. Additionally, reduced commuting time can result in more productive work hours. The implementation of the virtual workplace, when executed proficiently, has a favourable influence on productivity and job satisfaction. The utilisation of efficient communication channels to achieve optimal efficiency is an essential element within any organisation (Boughzala, de Vreede, & Limayem, 2012). For instance, the integration of diverse technological solutions, like video conferencing, project management software, and virtual workspaces, has significantly enhanced the effectiveness of cooperation within geographically dispersed teams. Based on the overhead discourse, it may be inferred that hypothesis three has been substantiated.

5 Conclusion

Several significant study findings have arisen because of the investigation of the use of the metaverse in the workplace. The effective use of the metaverse in the workplace has been found to have a favourable influence on both productivity and job satisfaction. The outcomes discussed in this context are heavily influenced by several key aspects, namely: effective communication channels, optimal utilisation of technology, establishment of trust, emphasis on achieving desired goals, and the cultivation of an inclusive culture. The implementation of the metaverse in the workplace offers employees the opportunity to exercise freedom in selecting their work environment and personalising their work schedules, thereby promoting work-life balance (Abdullah et al., 2022). The provision of flexibility in work arrangements facilitates the achievement of a more harmonious equilibrium between work and personal life. The integration of a metaverse inside the office environment has been shown to enhance collaboration and foster innovation. The utilisation of various technologies, including video conferencing, project management tools, and virtual workspaces, has facilitated efficient collaboration among remote teams. Consequently, this has resulted in enhanced capacities for innovation and problem-solving. Organisational emphasis on employee engagement and provision of a metaverse working experience are associated with increased employee retention rates. Organisations can enhance their ability to attract and retain high-performing individuals by offering freedom and autonomy.

6 Recommendations for Future Research

Considerable advancements have been achieved in comprehending the metaverse within the context of professional environments. However, there exist certain domains that necessitate additional analysis and exploration. Future research should investigate the influence of the metaverse on organisational culture within the workplace. Examining elements such as various leadership styles, effective communication strategies, and the dynamics within teams can yield valuable insights on the sustenance of a robust organisational culture within a metaverse workplace setting. The primary objective of research should be to ascertain optimal managerial strategies for effectively leading hybrid teams. The comprehension of proficient managerial tactics, encompassing performance assessment, personnel growth, and the cultivation of team unity, can facilitate organisations in maximising their utilisation of metaverse work arrangements. Long-Term Effects: Further research is needed to understand the long-term effects of the metaverse workplace on employee well-being, job satisfaction, and organizational outcomes. Longitudinal studies offer valuable insights into the long-term sustainability and longevity of the metaverse work environment. The research should aim to investigate the possible obstacles and remedies in guaranteeing equitable opportunities, resource accessibility, and inclusiveness for all employees, irrespective of their geographical work settings. Future research could explore the psychological and social aspects that exert effect on the experiences of employees in metaverse work environments. An examination of several elements, including motivation, social connection, and psychological well-being, can provide valuable insights into the development of methods aimed at supporting employees in the metaverse within the workplace.

Furthermore, it is advisable to cultivate a workplace atmosphere that promotes the establishment of explicit boundaries among employees, while concurrently providing resources and support mechanisms to effectively cope with stress and improve overall welfare. The metaverse paradigm has the potential to inadvertently exacerbate pre-existing inequalities within organisational hierarchies. Not all employees have equal access to remote employment prospects or the necessary resources to effectively participate in telecommuting. Organisations must place utmost importance on the principles of fairness, inclusivity, and accessibility while adopting the metaverse model, considering the varied needs and circumstances of their workforce (Dwivedi et al., 2023).

References

Abbas, S., et al.: Antecedents of trustworthiness of social commerce platforms: a case of rural communities using multi group SEM & MCDM methods. Electron. Commer. Res. Appl. **62**, 101322 (2023)

Abdullah, H.O., et al.: Predicting determinants of use mobile commerce through modelling non-linear relationships. Cent. Eur. Bus. Rev. **11**(5), 23 (2022)

Agho, A.O., Mueller, C.W., Price, J.L.: Determinants of employee job satisfaction: an empirical test of a causal model. Hum. Relat. **46**(8), 1007–1027 (1993)

Ahmad, M.B., Wasay, E., Ullah, S.: Impact of employee motivation on customer satisfaction: study of airline industry in Pakistan. Interdisc. J. Contemp. Res. Bus. **4**(6), 531–539 (2012)

Al-Abrrow, H., Alnoor, A., Ismail, E., Eneizan, B., Makhamreh, H.Z.: Psychological contract and organizational misbehavior: exploring the moderating and mediating effects of organizational health and psychological contract breach in Iraqi oil tanks company. Cogent Bus. Manag. **6**(1), 1683123 (2019)

Aldowah, H., Rehman, S.U., Ghazal, S., Umar, I.N.: Internet of things in higher education: a study on future learning. J. Phys. Conf. Ser. **892**(1), 12017 (2017)

Alfaisal, R., Hashim, H., Azizan, U.H.: Metaverse system adoption in education: a systematic literature review. J. Comput. Educ., 1–45 (2022)

AL-Fatlawey, M.H., Brias, A.K., Atiyah, A.G.: The role of strategic behavior in achievement the organizational excellence "analytical research of the manager's views of Ur state company at Thi-Qar governorate". J. Adm. Econ. **10**(37) (2021)

Alharbi, R., Alnoor, A.: The influence of emotional intelligence and personal styles of dealing with conflict on strategic decisions. PSU Res. Rev. (2022)

Al-Hchaimi, A.A.J., Sulaiman, N.B., Mustafa, M.A.B., Mohtar, M.N.B., Hassan, S.L.B.M., Muhsen, Y.R.: Evaluation approach for efficient countermeasure techniques against denial-of-service attack on MPSoC-based IoT using multi-criteria decision-making. IEEE Access **11**, 89–106 (2022)

Aliyyah, N., et al.: What affects employee performance through work motivation? J. Manag. Inf. Decis. Sci. **24**(1) (2021)

Alnoor, A., et al.: How positive and negative electronic word of mouth (eWOM) affects customers' intention to use social commerce? A dual-stage multi group-SEM and ANN analysis. Int. J. Hum. Comput. Interact., 1–30 (2022)

Alsalem, M.A.: Rescuing emergency cases of COVID-19 patients: an intelligent real-time MSC transfusion framework based on multicriteria decision-making methods. Appl. Intell., 1–25 (2022)

Arif, B., Farooqi, Y.A.: Impact of work life balance on job satisfaction and organizational commitment among university teachers: a case study of University of Gujrat, Pakistan. Int. J. Multidisc. Sci. Eng. **5**(9), 24–29 (2014)

Atiyah, A.G.: Effect of temporal and spatial myopia on managerial performance. Journal La Bisecoman **3**(4), 140–150 (2022)

Atiyah, A.G., Zaidan, R.A.: Barriers to using social commerce. In: Alnoor, A., Wah, K.K., Hassan, A. (eds.) Artificial Neural Networks and Structural Equation Modeling: Marketing and Consumer Research Applications, pp. 115–130. Springer, Singapore (2022). https://doi.org/10.1007/978-981-19-6509-8_7

Aziri, B.: Job satisfaction: a literature review. Manag. Res. Pract. **3**(4) (2011)

Bakotić, D.: Relationship between job satisfaction and organisational performance. Econ. Res.-Ekonomska Istrazivanja **29**(1), 118–130 (2016). https://doi.org/10.1080/1331677X.2016.1163946

Ball, M.: The Metaverse: and How It Will Revolutionize Everything. Liveright Publishing (2022)

Barry, D.M., et al.: International comparison for problem based learning in metaverse. In: The ICEE and ICEER, 6066 (2009)

Belwal, S., Belwal, R., Al-Hashemi, S.E.: Family friendly policies and the Omani Labour Law. Empl. Relat. Int. J. (2019)

Bhatti, K.K., Qureshi, T.M.: International review of business research papers impact of employee participation on job satisfaction, employee commitment and employee productivity. Hum. Resour. Manage. **3**(2), 54–68 (2007)

Boughzala, I., de Vreede, G.-J., Limayem, M.: Team collaboration in virtual worlds: editorial to the special issue. J. Assoc. Inf. Syst. **13**(10), 6 (2012)

Bozanic, D., Tešić, D., Puška, A., Štilić, A., Muhsen, Y.R.: Ranking challenges, risks and threats using fuzzy inference system. Decis. Making Appl. Manag. Eng. **6**(2), 933–947 (2023)

Chew, X., Khaw, K.W., Alnoor, A., Ferasso, M., Al Halbusi, H., Muhsen, Y.R.: Circular economy of medical waste: novel intelligent medical waste management framework based on extension linear Diophantine fuzzy FDOSM and neural network approach. Environ. Sci. Pollut. Res., 1–27 (2023)

Cho, H.Y., Lee, H.-J.: Digital transformation for efficient communication in the workplace: analyzing the flow coworking tool. Bus. Commun. Res. Pract. **5**(1), 20–28 (2022)

Clark, S.C.: Work/family border theory: a new theory of work/family balance. Hum. Relations **53**(6), 747–770 (2000)

Deshpande, B., Arekar, K., Sharma, R., Somaiya, S.: Effect of employee satisfaction on organization performance: an empirical study in hotel industry. In: Ninth AIMS International Conference on Management Held at Pune, India, January, pp. 1–4 (2012)

Dionisio, J.D.N., Iii, W.G.B., Gilbert, R.: 3D virtual worlds and the metaverse: current status and future possibilities. ACM Comput. Surv. (CSUR) **45**(3), 1–38 (2013)

Dwivedi, Y.K., et al.: So What if ChatGPT wrote it? multidisciplinary perspectives on opportunities, challenges and implications of generative conversational AI for research, practice and policy. Int. J. Inf. Manag. **71**, 102642 (2023)

Falchuk, B., Loeb, S., Neff, R.: The social metaverse: battle for privacy. IEEE Technol. Soc. Mag. **37**(2), 52–61 (2018)

Felstead, A., Jewson, N., Phizacklea, A., Walters, S.: Opportunities to work at home in the context of work-life balance. Hum. Resour. Manag. J. **12**(1), 54–76 (2002)

Gartner_Inc. (n.d.): What is a metaverse? Gartner. Retrieved October 3 (2023). https://www.gartner.co.uk/en/articles/what-is-a-metaverse

Garg, P., Yajurvedi, N.: Impact of work-life balance practices on employees retention and organizational performance–a study on IT industry. Indian J. Appl. Res. **6**(8), 105–108 (2018)

Gatea, A.A., Marina, V.: Higher education funding in Iraq in terms of the experience of particular developed countries. Int. J. Adv. Stud. **6**(1), 8–17 (2016)

Griol Barres, D., Sanchis de Miguel, M.A., Molina López, J.M., Callejas, Z.: Developing enhanced conversational agents for social virtual worlds (2019)

Güngör, P.: The relationship between reward management system and employee performance with the mediating role of motivation: a quantitative study on global banks. Procedia Soc. Behav. Sci. **24**, 1510–1520 (2011)

Gunlu, E., Aksarayli, M., Perçin, N.Ş.: Job satisfaction and organizational commitment of hotel managers in Turkey. Int. J. Contemp. Hospitality Manag. (2010)

Harhara, A.S., Singh, S.K., Hussain, M.: Correlates of employee turnover intentions in oil and gas industry in the UAE. Int. J. Organ. Anal. **23**(3), 493–504 (2015)

Hatane, S.E., Sondak, L., Tarigan, J., Kwistianus, H., Sany, S.: Eyeballing internal auditors' and the firms' intention to adopt Metaverse technologies: case study in Indonesia. J. Financ. Reporting Account. (2023)

Hopkins, J., Bardoel, A.: The future is hybrid: how organisations are designing and supporting sustainable hybrid work models in post-pandemic Australia. Sustainability **15**(4), 3086 (2023)

Hwang, G.-J., Chien, S.-Y.: Definition, roles, and potential research issues of the metaverse in education: an artificial intelligence perspective. Comput. Educ. Artif. Intell. **3**, 100082 (2022)

Hwang, I., Shim, H., Lee, W.J.: Do an organization's digital transformation and employees' digital competence catalyze the use of telepresence? Sustainability **14**(14), 8604 (2022)

Igbinomwanhia, O., Iyayi, O., Iyayi, F.: Employee work-life balance as an HR imperative. Afr. Res. Rev. **6**(3), 109–126 (2012). https://doi.org/10.4314/afrrev.v6i3.8

Judge, T.A., Thoresen, C.J., Bono, J.E., Patton, G.K.: The job satisfaction–job performance relationship: a qualitative and quantitative review. Psychol. Bull. **127**(3), 376 (2001)

Kaaria, A.G., Mwaruta, S.S.: Mental health ingenuities and the role of computer technology on employees' mental health: a systematic review. East Afr. J. Health Sci. **6**(1), 219–231 (2023)

Kanematsu, H., Kobayashi, T., Ogawa, N., Barry, D.M., Fukumura, Y., Nagai, H.: Eco car project for Japan students as a virtual PBL class. Procedia Comput. Sci. **22**, 828–835 (2013)

Kibui, A.W., Gachunga, H., Namusonge, G.S.: Role of talent management on employees retention in Kenya: a survey of state corporations in Kenya: empirical review. Int. J. Sci. Res. **3**(2), 414–424 (2014)

Kim, M., Knutson, B.J., Choi, L.: The effects of employee voice and delight on job satisfaction and behaviors: comparison between employee generations. J. Hosp. Market. Manag. **25**(5), 563–588 (2016)

Kohler, S.S., Mathieu, J.E.: An examination of the relationship between affective reactions, work perceptions, individual resource characteristics, and multiple absence criteria. J. Organ. Behav. **14**(6), 515–530 (1993)

Koohang, A., et al.: Shaping the metaverse into reality: a holistic multidisciplinary understanding of opportunities, challenges, and avenues for future investigation. J. Comput. Inf. Syst. **63**(3), 735–765 (2023)

Lazar, I., Osoian, C., Ratiu, P.: The role of work-life balance practices in order to improve organizational performance (2010)

Leitão, J., Pereira, D., Gonçalves, Â.: Quality of work life and organizational performance: workers' feelings of contributing, or not, to the organization's productivity. Int. J. Environ. Res. Public Health **16**(20), 3803 (2019)

Leners, D.W., Roehrs, C., Piccone, A.V.: Tracking the development of professional values in undergraduate nursing students. J. Nurs. Educ. **45**(12), 504–511 (2006)

Loan-Clarke, J., Arnold, J., Coombs, C., Hartley, R., Bosley, S.: Retention, turnover and return–a longitudinal study of allied health professionals in Britain. Hum. Resour. Manag. J. **20**(4), 391–406 (2010)

Manhal, M., Al-khalidi, A., Hamad, Z.: Strategic network: managerial myopia point of view. Manag. Sci. Lett. **13**(3), 211–218 (2023)

Maurer, M.: Accounting firms scoop up virtual land in the Metaverse. Wall Street J. (2022)

Muhsen, Y.R., et al.: Enhancing NoC-based MPSoC performance: a predictive approach with ANN and guaranteed convergence arithmetic optimization algorithm. IEEE Access (2023)

Ojo, I.S., Salau, O.P., Falola, H.O.: Work-life balance practices in Nigeria: a comparison of three sectors. **6**(2), 3–14 (2014)

Ondrejka, C.: Escaping the gilded cage: user created content and building the metaverse. NYLS Law Rev. **49**(1), 6 (2005)

Organ, D.W.: Organizational Citizenship Behavior: The Good Soldier Syndrome. Lexington Books/DC Heath and Com (1988)

Paais, M., Pattiruhu, J.R.: Effect of motivation, leadership, and organizational culture on satisfaction and employee performance. J. Asian Financ. Econ. Bus. **7**(8), 577–588 (2020)

Pandey, C., Khare, R.: Impact of job satisfaction and organizational commitment on employee loyalty. Int. J. Soc. Sci. Interdisc. Res. **1**(8), 26–41 (2012)

Park, H., Ahn, D., Lee, J.: Towards a metaverse workspace: opportunities, challenges, and design implications. In: Proceedings of the 2023 CHI Conference on Human Factors in Computing Systems, pp. 1–20 (2023)

Patriota, D.: Employee Retention: An Integrative View of Supportive Human Resource Practices and Perceived Organizational Support (2009)

Peters, P., Heusinkveld, S.: Institutional explanations for managers' attitudes towards telehomeworking. Hum. Relat. **63**(1), 107–135 (2010)

Potkány, M., Giertl, G.: Statistical prognosis of basic business performance indicators in the wood processing industry of the Slovac Republic. In: Markets for Wood and Wooden Products, Zagreb: WoodEMA, pp. 31–56 (2013)

Pushpakumari, M.D.: The impact of job satisfaction on job performance: an empirical analysis. City Forum **9**(1), 89–105 (2008)

Recker, J.C., Lukyanenko, R., Jabbari Sabegh, M., Samuel, B., Castellanos, A.: From representation to mediation: a new agenda for conceptual modeling research in a digital world. MIS Q. Manag. Inf. Syst. **45**(1), 269–300 (2021)

Rizwan, M., Tariq, M., Hassan, R., Sultan, A.: A comparative analysis of the factors effecting the employee motivation and employee performance in Pakistan. Int. J. Hum. Resour. Stud. **4**(3), 35 (2014)

Saari, L.M., Judge, T.A.: Employee attitudes and job satisfaction. Human Resource Management: Published in Cooperation with the School of Business Administration, The University of Michigan and in Alliance with the Society of Human Resources Management **43**(4), 395–407 (2004)

Sadaa, A.M., et al.: Based on the perception of ethics in social commerce platforms: adopting SEM and MCDM approaches for benchmarking customers in rural communities. Current Psychol., 1–35 (2022)

Sari, E.T.: Motivation and satisfaction towards employees' loyality to achive company's advantage. JMBI UNSRAT (Jurnal Ilmiah Manajemen Bisnis Dan Inovasi Universitas Sam Ratulangi) **4**(1) (2018)

Shahzadi, I., Javed, A., Pirzada, S.S., Nasreen, S., Khanam, F.: Impact of employee motivation on employee performance. Eur. J. Bus. Manag. **6**(23), 159–166 (2014)

Shravanthi, A.R., Deshmukh, S., Deepa, N.: Work life balance of women in India work life balance of women in India. Int. J. Res. Manag. Sci. **1**(1), 47–56 (2015)

Smith, P.C., Kendall, L.M., Hulin, C.L.: The measurement of satisfaction in work and retirement: a strategy for the study of attitudes (1969)

Weiss, H.M.: Deconstructing job satisfaction: separating evaluations, beliefs and affective experiences. Hum. Resour. Manag. Rev. **12**(2), 173–194 (2002)

Wilkinson, A.: The other side of quality: 'soft' issues and the human resource dimension. Total Qual. Manag. **3**(3), 323–330 (1992)

Wu, D.-Y., Lin, J.-H.T., Bowman, N.D.: Watching VR advertising together: how 3D animated agents influence audience responses and enjoyment to VR advertising. Comput. Hum. Behav. **133**, 107255 (2022)

Yang, H.: Efficiency wages and subjective performance pay. Econ. Inq. **46**(2), 179–196 (2008)

Yemenici, A.D.: Entrepreneurship in the world of metaverse: virtual or real? J. Metaverse **2**(2), 71–82 (2022)

Measuring the Possibility of Adopting Metaverse Technology as an Appropriate Strategy to Achieve a Sustainable Competitive Advantage

Hashim Nayef Hashim Al-Hachim(✉) and Adnan Saad Tuama Al-Sukaini

Management Technical College, Southern Technical University, Basra, Iraq
{hashim.naif,adnan.saad}@stu.edu.iq

Abstract. The metaverse is a digital transformation that represents the creation of an incredible atmosphere for users who often rely on virtual reality or augmented reality to interact with the surrounding environment. Despite the amount of knowledge and awareness required to adopt this technology, as well as the high costs, the benefit it achieves accrues to all parties of interest it represents an incentive that encourages the adoption of this technology as a business strategy. Accordingly, this study aims to determine the extent of the possibility of using metaverse technology as a strong business strategy leading to achieving a sustainable competitive advantage for the organization. Data was collected from industrial companies in Basra, and the number of respondents reached 100 people. Using PLS-SEM the results of the study showed the possibility of using this metaverse technology as a business strategy that contributes to achieving a sustainable competitive advantage in the company sample of the study.

Keywords: Metaverse · Sustainable competitive advantage · Virtual reality

1 Introduction

With the development of the digital landscape, the metaverse appears as a trend to change the rules of competition in all businesses at all levels. The metaverse is defined as "a large-scale standard that works simultaneously and continuously by an unlimited number of many users with the effectiveness of an individual sense of existence" (Tan et al. 2023). However, given the experience, which is characterized by its intangibility and the possibility of showing its results based on multi-sensory and multifaceted aspects, it can be said that it is not possible to rely on a specific measure to know the final impact of the metaverse strategy (Krishnamurthy et al. 2022). Metaverse describes everything experienced with the Internet today. The metaverse helps to move from one world to another so that we can work, play, and communicate in the best way possible. Many companies have introduced metaverse technology as their work context. More effort is being made to introduce more improvements to it and benefit from it in the greatest possible way (Farhi et al. 2023). Therefore, the metaverse represents one of the most developed technologies at the present time and depends on the availability of a fixed, virtual, three-dimensional environment. Individuals can communicate and participate with

M. Al-Emran et al. (Eds.): IMDC-IST 2024, LNNS 876, pp. 69–82, 2023.
https://doi.org/10.1007/978-3-031-51300-8_5

others, and practitioners and researchers believe that this technology has the potential to state most of the negative effects resulting from social distancing, and most importantly, it represents a completely new communication revolution (Di Dario et al. 2022).

The metaverse can be described as a computer implementation of a large-scale universe that, in one way or another, supports relevant or entertaining applications within that universe. For the sake of simplicity, it must be taken into account that each universe within the real multiverse includes a specific application domain, with conceptual specifications that distinguish it from others. The features of the other universe, as these specifications must be subject to a set of rules, called integrity constraints in database language, the purpose of which is to control the behavior of the states composing each of these fields (Furtado et al. 2023). Finally, it must be pointed out that the metaverse is strongly linked to the real world and has a real identity, and it is likely that there will be many companies working together to create one large metaverse in the future. Then these companies need to make significant investments to achieve an acceptable level of privacy, and metaverse security (Yemenici 2022). Therefore, companies that want to strengthen their position in the global competitive market can adopt a metaverse strategy to achieve sustainability in their business and within the context in which they operate. To this end, this study aims to investigate the role of virtual reality in achieving competitive advantage.

2 Literature Review

2.1 Metaverse

Metaverse is a concept that refers to a new beginning to create something new, somewhat similar to the early days of the Internet. The concept of the metaverse was coined in 1992 by science fiction creator Neal Stevenson. He basically describes it as “a virtual world in which people come together to work, play, and socialize.” " It is a simulated digital environment that combines augmented reality (AR) and virtual reality (VR) to create areas for users to interact in a simulated way with the real world, (Laeeq 2022). It is expected that the development of the metaverse will be accompanied by the transformation of imagination into reality by creating a form of convergence of different technologies, which can be described as a means of sustainable education, stripped of the constraints of time and space. At the beginning of its emergence, the metaverse referred to a real world linked in a way that makes one act as an alternative to oneself in augmented reality. The level of interaction and actual participation has also become possible thanks to the level achieved electronically, as individuals, through this technology, have new identities and distinct capabilities similar to those they possess in the real world, (Singh et al. 2022). The term “metaverse” is constantly evolving due to continuous technical progress that contributes to changing the nature of interaction between individuals, and this technology, in turn, has captured the collective consciousness of researchers and the rest of society in general, making it a fascinating research focus (Krishnasamy et al. 2023). By the year 2026, it is likely that 25% of people will devote at least an hour of their time daily to the Metaverse to address various purposes such as learning, shopping, work, entertainment, and social networking. Metaverse also enables users to have meaningful conversations with their friends as well as shop online (Chandiwala et al. 2023).

The level of reliability of the Metaverse can be measured by its level of security to know which systems enable access only to authorized persons while providing protection against the malicious activities of hackers (Muhsen et al. 2023; Al-Hchaimi et al. 2022; Chew et al. 2023). In addition, access to data inventory is important to provide the advantage of simplicity for individuals to access their digital data via the Metaverse. It is worth noting here that individuals need Internet of Things devices to be able to experience sensations within virtual worlds (Sami et al. 2023). Finally, Metaverse is a technological transformation that creates an incredible atmosphere for users, who often rely on VR or AR to interact with the surrounding environment. Unlike some additional digital rules, the metaverse is a pervasive technology that has the ability to live on in reality beyond the necessity of being "on" or "off" in order to maintain continuous functioning or operation (Singh et al. 2022).

2.2 The Development of Metaverse Technology

The term metaverse has been used since at least 1992 and currently refers generally to the concept of a virtual and persistent world through which users are able to communicate and interact with other users and participate in social activities within the surrounding environment, more or less similar to the interactions of the physical world. The term "metaverse" refers to a general form of associated technologies and not to a specific configuration of devices, applications, services, and specific platforms (Furtado et al. 2023). Therefore, it can be said that the concept of the Metaverse originally existed for decades, but it gained global attention coinciding with the renaming of the brand "Meta." Recent developments in computing technologies have also contributed to the opportunity to accelerate its way of thinking and create a global computing network called the "Metaverse." Metaverse is characterized as a development of the Internet today, through which people can obtain virtual identities and presence, interact from one party to another, and complete and create multiparty transactions (Cho 2023).

The first researchers who dealt with the metaverse suggested that it is based on five basic dimensions: 1- The design of the metaverse 2- The way in which people are represented in the virtual world 3- The capabilities that enable the metaverse to communicate, interact and display 4- The behaviors that can be represented via Symbolic images 5- Results, which in turn indicate the effectiveness of cooperation (Lembacher 2022). Khan et al. (2022) identified three basic steps for developing metaverse technology. The first step is to develop a real model (i.e., digital representation) of physical reality. This modeling can be achieved using various means, including, for example, mathematical modeling, data-based modeling, simulation, and experimental modeling. Second, additional information (such as virtual objects for mobile environments) is imposed on the virtual model through the use of various monitoring and sensing techniques. The third and final step is to create digital avatars of individuals in the metaverse that can represent a variety of stakeholders such as network operators and users (Abbas et al. 2023; Bozanic et al. 2023). The metaverse is also distinguished by its unique features that enable companies to transfer their message, values, and vision to a digital environment. It must be noted that there are five basic steps that these companies must adopt in order to adopt the concept of the metaverse: exploration, assurance, activation, co-creation, and communication (Dima and Vargas 2021). It is also possible to classify four basic metaverse technologies: 1- Virtual world technologies, which in turn affect the general life of the physical world. 2- Technologies enhanced with information that reflects the real world. 3- AR technologies, which in turn support the individual's actual world. 4- The real record, which in turn is responsible for recording various states such as emotion (Cho 2023). Abovitz et al. (2022) stated in their research paper regarding building a metaverse strategy based on ten main steps, as shown in Fig. 1 and Table 1.

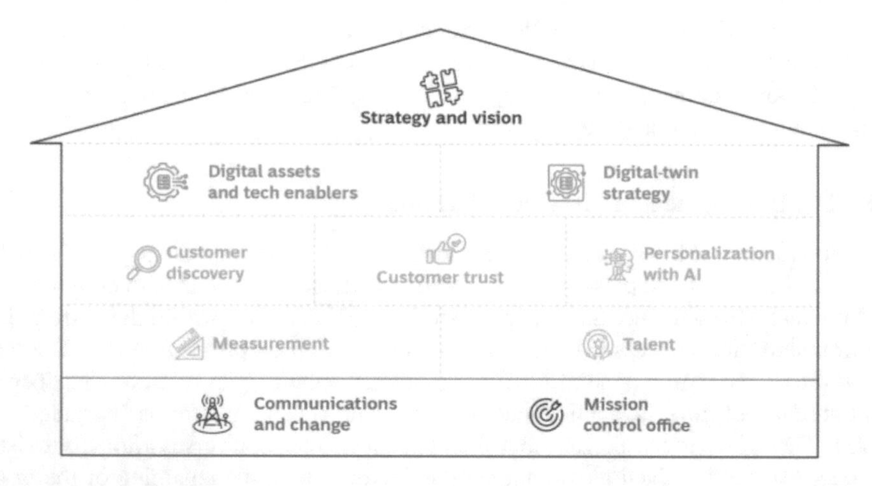

Fig. 1. The proposed ten steps for building a metaverse strategy. 2022 **Source:** Author adoption from Abovitz et al. (). How the metaverse will remake your strategy. Boston Consulting Group, 1(July).

Table 1. Terms and concepts related to the search.

	Term	Concept	reference
1	Vision	The organization's vision reflects the desired future state	(Tanković 2013)
2	Strategy	It consists of a set of goals and important information to achieve these goals, who will do them, how to do them, for whom, and why the results of these goals are valuable	(Tanković 2013)

(*continued*)

Table 1. (*continued*)

	Term	Concept	reference
3	Digital assets	They are intangible assets, not paper money. They also work as a means of exchange, as some have begun to achieve great success through them as a strategy for alternative investments	(Merwe 2021)
4	Digital twin	It is a realistic digital representation of processes, assets and systems in the built or natural environment	(Lu et al. 2020)
5	Customer discovery	A process through which business leaders build their hypotheses about their business models and then attempt to verify or deny the validity of these hypotheses by conducting various interviews with customers	(Batova et al. 2016)
6	Customer trust	It is the amount of positive expectation and willingness that a customer makes to buy a particular product	(Manzoor et al. 2020)
7	Measurement	The measure is a statistical model that links unobservable theoretical blocks and is represented as latent variables with unobservable characteristics	(Jacobs and Wallach 2021)
8	Communication	It is one of the most attractive forms of interaction in a way that combines advanced technologies such as augmented reality and virtual reality with the aim of creating rich, multi-characteristic experiences in a way that simulates the interaction achieved in the real world	(Ashraf et al. 2023)
9	Talent	A combination of capabilities developed by people in an organization that aim to achieve a set of advantages in the organization in which they work	(Kravariti and Johnston 2019)
10	Mission	The mission explains the reason and purpose of the organization's current existence	(Tanković 2013)

2.3 Importance of Metaverse

Today, many international companies use VR and AR technologies to improve their relationship with their customers and provide them with experiences that could not be done previously. An example of this is what the vehicle manufacturing company (Ferrari) is doing for its new model in augmented reality, as buyers can't only see the car, but it has also become possible to experience the brakes and mechanical systems of the car through virtual reality (Farhi et al. 2023). Therefore, it is right for companies to work on delving into the details of the metaverse to learn about the advantages of this advanced technology. Such companies can expand the customer base is that the metaverse knows no limits of place or time. In addition to the possibility of providing a rare dynamic environment for interacting with geographically diverse environments in ways that are no longer feasible in the real world (Chandiwala et al. 2023). In contrast to the traditional experiences of the online purchasing process, companies in the shadow of the Metaverse are now able to provide their customers with an interactive environment through which they can feel real emotions related to the products that are displayed. Companies are now able to make customers' focus on the things that are dynamically displayed in an interactive situation (Singh et al. 2022). Thus, the Metaverse narrows down many parts of the customer journey, valuing consumers' VR prowess and collaborating with technology vendors to design the Metaverse's ambience and providing the necessary accessories in order to make the journey more connected, seamless, and engaging, Meaning (Krishnamurthy et al. 2022).

The metaverse can be described as a large digital environment in which people interact in real time with the same emotions they would feel if they were communicating with each other in the real world. The metaverse extends as far as our imagination allows and can be considered endless as long as it is not bound by the constraints of time or space. Users in the Metaverse will be able to communicate with each other at an ideal time (Singla and Himangi 2023). Even if the Metaverse fails to represent the dazzling visions created by those interested in science fiction, it will generate a significant amount of value as a new computing platform or medium for creating content. Metaverse represents the gateway to most digital experiments, a vital component of all scientific experiments, and an important business platform. For this reason, the Metaverse has the potential to replace the Internet and show significant growth in economic terms. With the Metaverse, new services, products, and companies will emerge to process payment, advertising, recruitment, and content creation. This naturally means that many of today's employees will be pushed out of their jobs (Yemenici 2022).

2.4 Sustainable Competitive Advantage

Competitive business world all organizations need to have the appropriate facilities to achieve distinctive results and thus outperform others and implementing a value creation strategy that no other organization adopts at the same time help to have a competitive advantage (Hamadamin and Atan 2019). When the market environment changes rapidly and competitive advantages are distinctly unsustainable, it is necessary to search for a supplier, which in turn represents a useful construct for understanding the ability of companies to achieve great performance results at a time when others fail. Thus, competitive advantage refers to "the ability of companies to achieve better performance." (Pratono et al. 2019). Sustainable competitive advantage is the most common concept in the field of management, as it explains the factors that affect performance across companies. This sustainable competitive advantage is achieved through the effective implementation of strategies that affect the overall activities of the organization. The organization needs practical innovations to maintain its competitive balance and ensure success (Arsawan et al. 2022). There are five main dimensions of sustainable competitive advantage. Flexibility in its simplest form can be described as the organization's high ability to deploy its resources with high speed, efficiency, and acceptable effectiveness in response to sudden changes (Dubey et al. 2021). Therefore, flexibility refers to the company's ability to absorb situations of turmoil and uncertainty and thus provide opportunities to achieve a sustainable competitive advantage for the organization (Yousuf et al. 2019). Cost, which can be considered a good step, at the same time companies need to carefully identify and evaluate the most important costs they seek to reduce (Dagnino et al. 2021). Therefore, it can be said that the low-cost advantage is based on activities based primarily on efficiency, and therefore the customer looks at the company's offers by comparing them with the offers of its competitors to estimate the potential value and choose the bids with the lowest value (Kaleka and Morgan 2017). Delivery and speed of delivery is one of the most important factors within the production process at the present time, especially since the production process today is based on timely production. Therefore, speedy delivery is an important element for increasing the competitiveness of the organization along with flexibility, quality and low cost (Kaewchur 2021). Quality is an important element that enables the organization to achieve sustainable competitive advantage. It is a strategic consideration because of its role in customer satisfaction. Many companies realize that they can compete by distinguishing themselves by improving their level of service as well as the quality.

2.5 Metaverse and Sustainable Competitive Advantage

Companies looking to compete must first understand how their strategy and sustainable advantages change as the competition evolves. This will certainly require them to evaluate what assets and capabilities they need to provide to adopt the metaverse, how to develop them and potential changes in their current value chain, and how they can defend their current position. These need companies to issue cases such as protecting the digital identity of their customers and providing links across data layers to enhance communication services (Abovitz et al. 2022). Whether the metaverse is created as an ecosystem under which the company operates to provide its services, this is very important from the perspective of the company's future strategy, as it provides a business ecosystem, achieves balanced regional development, and creates job opportunities by developing an entrepreneurial economy that enhances social and global well-being, thus achieving sustainable growth in the future. To create the technological paradigm represented by the metaverse, it is necessary to promote appropriate strategies and policies (Wang et al 2022). The metaverse can be adopted as a business model. It covers many industries and has rich connotations and is characterized by the comprehensiveness of customer segmentation, value proposition, channels, customer relationships, key partners, key resources, key activities, cost structure, and revenue streams. It is a model it is built on the basis of covering all elements of value, which facilitates excellence in work (Tallón and Santana 2023). People behave differently when accepting or rejecting new technologies, and therefore two tendencies appear in accepting or rejecting any new technology: "Technophilic," which is defined as accepting new technology. And "Technophobia," which is defined as a reluctance to accept new technology. Here it is necessary It is noteworthy that despite the shortcomings of the Metaverse, like any other technology, it may create an impact due to current and future technological progress as well. Finally, it can be said that the Metaverse is a promising technology that individuals and companies alike need to ensure their continued control over the business (Ganapuram et al. 2022). Thus, we assume that:

There is a relationship between metaverse adoption and sustainable competitive advantage.

3 Methodology

The questionnaire was measured on a scale of 1 to 5, ranging from strongly agree to strongly disagree, respectively. These items were related to the five main variables of the study. The responses were planned for data adequacy analyses and correlation studies using the Kaiser-Meyer-Olkin adequacy scale. Sampling (KMO) and Bartlett's test. Also perform principal components analysis (PCA) to check the population and variance of the data (Table 2).

Table 2. Variables and classification.

Variable No	Variables	Classification
1	Benefit for users	Facilities
2	Profit for producers	
3	Ease of use of the Metaverse	
4	Knowledge and awareness required to develop the Metaverse	
5	Costs required to develop Metaverse technology	Difficulties

Table 3 indicates that the majority of the sample members were male, at a rate of 79%. 44% of participants reported that their ages ranged between 40 and 49 years. As for the level of knowledge of respondents regarding metaverse technology, 44% of the sample had an average level of knowledge related to metaverse technology. The number of years of job service differed between participants. The number of years of job service differed among the sample members. There were 33% of participants who had 6–10 years of experience, 27% had 11–15 years of experience, and 40% had 3–5 years of experience.

As mentioned in the previous section, data are collected from different age groups and genders, with diverse levels of knowledge. Most of the desirable people are male and their ages range between 40–49 and most respondents have moderate knowledge related to the Metaverse.

Table 3. Demographic characteristics of respondents.

	Freq		%		Freq	%
Gender				**Age**		
Male	47		79	**19–29**	11.4	19
Female	13		21	**30–39**	9.6	16
Total	**60**		**100**	**40–49**	26.4	44
				50and more	12.6	21
				Total	**60**	**100**
level of knowledge regarding metaverse technology				**Job Tenure**		
Low level		6.6	11	**Less Than 3 Years**	7	11.7
Acceptable level		12.6	21	**3–5**	11	18.3
Medium level		26.4	44	**6–10**	11	18.3
High level		11.4	19	**11–15**	12	20
Advanced level		3	5	**More Than 15**	19	31.7
Total		60	100	**Total**	**60**	**100**

4 Results and Analysis

The questionnaire items that appear in Tables 4 and 5 were formulated on the basis of what was stated by (Al-Rumaidi 2019) and (Khatolwa et al. 2018) where Table 4 shows the means and standard deviation for all metaverse elements as a strategy that can be adopted. The average answer score for the sample members regarding the independent variable represented by the metaverse is 3.31, with a standard deviation of 0.84. This result means the industrial companies have the basic understanding that enables it to adopt the metaverse as a business strategy, enabling it to adapt to the changing business environment, with the least possible and provide high quality products that are flexible with this environment.

Table 4. Mean rating of metaverse.

Items	Mean	SD
The organization continually seeks to improve its competitive position compared to its competitors in the future	3.11	1.13
The organization develops its own strategy taking into account developments in the environment in which it operates	3.10	1.21

(*continued*)

Table 4. (*continued*)

Items	Mean	SD
The organization has a sufficient budget to enable it to develop or acquire new digital assets	3.21	1.09
The organization has formal mechanisms to link its digital assets with other assets, and therefore it enjoys distinct levels of digital twinning	3.24	0.95
The organization allocates a lot of time, money and other resources to discovering new customers	3.23	0.92
The organization regularly reviews the level of satisfaction of its customers with the aim of verifying their level of trust	3.45	0.75
organization shares financial and business strategy information with all employees	3.88	1.25
The organization encourages managers to develop the leadership talents of their direct reports	3.42	1.03
The organization has a unified purpose/mission, and does not seek only profitability and growth	3.25	1.05
Total mean/standard deviation of Metaverse	3.31	0.84

Table 5 shows the means and standard deviation for all elements of sustainable competitive advantage. The average competitive advantage score is 3.21, with a standard deviation of 0.87. The average score for the flexibility element achieved an average of 3.13, with a standard deviation of 0.96, the cost element achieved an average of 3.31, with a standard deviation of 0.90, the delivery speed element achieved an average of 3.27, with a standard deviation of 1.26, while the product quality element achieved an average of 3.12, with a standard deviation of 0.97.

Table 5. Mean rating of sustainable competitive advantage.

Items	Mean	SD
Flexibility		
The organization has a high ability to change its sales mix compared to its competitors	3.21	1.14
The organization has the ability to respond to changes in customer tastes	3.17	1.12
The organization is characterized by the ability to quickly change the size of the products it offers to its customers compared to its competitors	3.02	1.21
Total mean / standard	3.13	.96

(*continued*)

Table 5. (*continued*)

Items	Mean	SD
Cost		
The organization has the ability to provide products at low prices compared to competitors	3.36	.87
The organization is distinguished by its high ability to provide its services at the lowest possible cost compared to its competitors	3.27	.99
The organization is characterized by having lower costs compared to competitors	3.29	.96
Total mean/standard	3.31	.90
Delivery Speed		
The organization is distinguished by its rapid and appropriate response to customer complaints compared to competitors	3.20	1.13
The organization has the ability to quickly change the type of services or the size of the products it offers to its customers compared to its competitors	3.28	1.02
The organization is characterized by the ability to change its sales mix quickly compared to its competitors	3.26	.99
Total mean/standard	3.27	1.26
Quality		
The organization is distinguished by providing products of high quality and reliability compared to its competitors	3.11	1.21
The organization is characterized by providing its products with a good level of service in a reliable manner compared to competitors	3.08	1.31
The organization is distinguished by providing products with an acceptable level of quality as a minimum compared to competitors	3.16	1.10
Total mean/standard	3.12	.97

5 Conclusions

The present research discussed the need to find a new technology that would achieve competitive sustainability for businesses in this competitive world. It also suggests that the Metaverse is the most recommended technology in this regard because it is one of the upcoming technologies that affect performance as a new business strategy providing a positive return to users as well as producers. Based on the results of the analysis of the research data, it has become clear that users are not yet well acquainted with the Metaverse, and therefore the responses were neutral in this regard to some questions. From the responses, it can be inferred that Metaverse, being a new technology, needs to develop the level of knowledge and create awareness among stakeholders in order to implement and facilitate the process of adopting this technology as a business strategy. It can also be concluded that such a technology requires high development costs, and in order for it to succeed, it requires the need to make an effort. More efforts should be made to reduce the costs of adopting this technology in a way that facilitates the process of adopting this technology before everyone else. As a result of the above, the present study encourages us to adopt the metaverse technology as a business strategy that helps the organization achieve excellent and competitive sustainability in the environment in which it operates. As is clear from the previous analysis, this technology brings a lot of benefit to both users and producers alike, in addition to the ease of adopting this technology. It must be noted that despite the somewhat high costs of use and development of this technology and the low level of awareness and knowledge among stakeholders regarding the metaverse, this does not prevent companies from adopting this technology.

References

Abbas, S., et al.: Antecedents of trustworthiness of social commerce platforms: a case of rural communities using multi group SEM & MCDM methods. Electron. Commerce Res. Appl. **62**, 101322 (2023)

Abovitz, R., et al.: How the metaverse will remake your strategy. Boston Consult. Group **1**(July) (2022)

Al-Hchaimi, A.A.J., Sulaiman, N.B., Mustafa, M.A.B., Mohtar, M.N.B., Hassan, S.L.B.M., Muhsen, Y.R.: Evaluation approach for efficient countermeasure techniques against denial-of-service attack on MPSoC-based IoT using multi-criteria decision-making. IEEE Access **11**, 89–106 (2022)

Al-Romeedy, B.S.: Strategic agility as a competitive advantage in airlines–case study: Egypt air. J. Faculty Tour. Hotels-Univ. Sadat City **3**(1), 1–15 (2019)

Arsawan, I.W.E., Koval, V., Rajiani, I., Rustiarini, N.W., Supartha, W.G., Suryantini, N.P.S.: Leveraging knowledge sharing and innovation culture into SMEs sustainable competitive advantage. Int. J. Product. Perform. Manag.Manag. **71**(2), 405–428 (2022)

Ashraf, M., Gaydamaka, A., Moltchanov, D., Mohammad, A., Tan, B.: Maximizing detection of target with multiple direction possibilities to support immersive communications in Metaverse. In: Proceedings of the 2nd Workshop on Integrated Sensing and Communications for Metaverse, pp. 30–35 (2023)

Batova, T., Clark, D., Card, D.: Challenges of lean customer discovery as invention. In: 2016 IEEE International Professional Communication Conference (IPCC), pp. 1–5. IEEE (2016)

Bozanic, D., Tešić, D., Puška, A., Štilić, A., Muhsen, Y.R.: Ranking challenges, risks and threats using fuzzy inference system. Dec. Mak. Appl. Manage. Eng. **6**(2), 933–947 (2023)

Chandiwala, M., Patel, P., Mehta, A.: Advertising and branding with metaverse. Dogo Rang. Res. J. **13**(6), 95–99 (2023)

Chew, X., et al.: Circular economy of medical waste: novel intelligent medical waste management framework based on extension linear Diophantine fuzzy FDOSM and neural network approach. Environ. Sci. Poll. Res. **30**, 60473–60499 (2023)

Cho, Y.C.: Metaverse for marketing in the public sector: implications on citizen relationship management. Korean J. Artif. Intell. **11**(2), 29–38 (2023)

Dagnino, G.B., Picone, P.M., Ferrigno, G.: Temporary competitive advantage: a state-of-the-art literature review and research directions. Int. J. Manag. Rev.Manag. Rev. **23**(1), 85–115 (2021)

Darmawan, D., Grenier, E.: Competitive advantage and service marketing mix. J. Soc. Sci. Stud. (JOS3) **1**(2), 75–80 (2021)

Di Dario, D., Bilotti, U., Sibilio, M., Gravino, C., Palomba, F. Toward a Secure Educational Metaverse: A Tale of Blockchain Design for Educational Environments.

Dima, A. (ed.): Resilience and economic intelligence through digitalization and big data analytics. De Gruyter (2021)

Dubey, R., Gunasekaran, A., Childe, S.J., Fosso Wamba, S., Roubaud, D., Foropon, C.: Empirical investigation of data analytics capability and organizational flexibility as complements to supply chain resilience. Int. J. Prod. Res. **59**(1), 110–128 (2021)

Farhi, F., Jeljeli, R., Zamoum, K., Boudhane, Y., Lagha, F.B.: Metaverse technology in communication practices: a case study of IT products retailers in the UAE. Emerg. Sci. J. **7**(3), 928–942 (2023)

Furtado, A.L., Casanova, M.A., de Lima, E.S.: Some Preliminary Steps Towards Metaverse Logic. arXiv preprint arXiv:2307.05574 (2023)

Ganapuram, V., Mangu, S., Prasad, M.V.: Business Sustainability Through Futuristic Strategy: Metaverse a Promising Technology. SSRN 4513020 (2023)

Hamadamin, H.H., Atan, T.: The impact of strategic human resource management practices on competitive advantage sustainability: the mediation of human capital development and employee commitment. Sustainability **11**(20), 5782 (2019)

Jacobs, A.Z., Wallach, H.: Measurement and fairness. In: Proceedings of the 2021 ACM Conference on Fairness, Accountability, and Transparency, pp. 375–385 (2021)

Kaewchur, P.: Role of inventory management on competitive advantage of small and medium companies in Thailand. Turkish J. Comput. Mathe. Educ. (TURCOMAT) **12**(8), 2753–2759 (2021)

Kaleka, A., Morgan, N.A.: Which competitive advantage (s)? Competitive advantage–market performance relationships in international markets. J. Int. Mark. **25**(4), 25–49 (2017)

Khan, L.U., Han, Z., Niyato, D., Guizani, M., Hong, C.S.: Metaverse for wireless systems: vision, enablers, architecture, and future directions. arXiv preprint arXiv:2207.00413 (2022).

Khatolwa, A.K.: Operations Strategies and Competitiveness of Kenyan Cooperative Sector (Doctoral dissertation, university of nairobi) (2018)

Kravariti, F., Johnston, K.: Talent management: a critical literature review and research agenda for public sector human resource management. Public Manag. Rev.Manag. Rev. **22**(1), 75–95 (2020)

Krishnamurthy, R., Chawla, V., Venkatramani, A., Jayan, G.: Transforming your brand using the metaverse: eight strategic elements to plan for. California Review Management (2022)

Krishnasamy, R., Vistisen, P., Nikolic, L. T., Hemmingsen, L., Scarpelli, M.: The Meta-stage: utilizing metaverse-enabling technologies for hybrid co-presence experiences. In: 4th EAI International Conference on Technology, Innovation, Entrepreneurship and Education. Springer (2023)

Laeeq, K.: Metaverse: why, how and what. How and What (2022)

Lembacher, S.: High fashion in virtual reality: is there competitive advantage from early entry into the metaverse? (Doctoral dissertation) (2022)

Lu, Q., Xie, X., Heaton, J., Parlikad, A.K., Schooling, J.: From BIM towards digital twin: strategy and future development for smart asset management. In: Borangiu, T., Trentesaux, D., Leitão, P., GiretBoggino, A., Botti, V. (eds.) SOHOMA 2019. SCI, vol. 853, pp. 392–404. Springer, Cham (2020). https://doi.org/10.1007/978-3-030-27477-1_30

Manzoor, U., Baig, S.A., Hashim, M., Sami, A.: Impact of social media marketing on consumer's purchase intentions: the mediating role of customer trust. Int. J. Entrep. Res. **3**(2), 41–48 (2020)

Muhsen, Y.R., et al.: Enhancing NoC-based MPSoC performance: a predictive approach with ANN and guaranteed convergence arithmetic optimization algorithm. IEEE Access (2023)

Pratono, A.H., Darmasetiawan, N.K., Yudiarso, A., Jeong, B.G.: Achieving sustainable competitive advantage through green entrepreneurial orientation and market orientation: the role of inter-organizational learning. Bottom Line **32**(1), 2–15 (2019)

Sami, H., et al.: The metaverse: survey, trends, novel pipeline ecosystem & future directions. arXiv preprint arXiv:2304.09240 (2023)

Singh, H., Gupta, S., Hassan, M.: Metaverse: a new digital marketing trend. Int. J. Multidisci. Res. Anal. **5**, 3623–3628 (2022)

Singla, M.: To enhance object detection speed in meta-verse using image processing and deep learning. Int. J. Intell. Syst. Appl. Eng. **11**(9s), 176–184 (2023)

Tallón-Ballesteros, A.J., Santana-Morales, P.: The theoretical basis and landing strategy of the metaverse business model. Digit. Manage. Innov. Proc. DMI **2022**(367), 100 (2023)

Tan, G.W.H., et al.: Metaverse in marketing and logistics: the state of the art and the path forward. Asia Pacific J. Mark. Logist. **35**, 2932–2946 (2023)

Tanković, A.Č: Defining strategy using vision and mission statements of Croatian organizations in times of crisis. Econ. Res.-Ekonomska istraživanja **26**(sup1), 331–342 (2013)

Van der Merwe, A.: A taxonomy of cryptocurrencies and other digital assets. Rev. Bus. **41**(1), 30–43 (2021)

Wang, F.Y., Qin, R., Wang, X., Hu, B.: Metasocieties in metaverse: metaeconomics and metamanagement for metaenterprises and metacities. IEEE Trans. Comput. Soc. Syst. **9**(1), 2–7 (2022)

Yemenici, A.D.: Entrepreneurship in the world of metaverse: virtual or real? J. Metaverse **2**(2), 71–82 (2022)

Yousuf, A., et al.: The effect of operational flexibility on performance: a field study on small and medium-sized industrial companies in Jordan (2019)

Unveiling the Quality Perception of Productivity from the Senses of Real-Time Multisensory Social Interactions Strategies in Metaverse

Abbas Gatea Atiyah(✉)

College of Administration and Economic, University of Thi-Qar, Thi-Qar, Iraq
abbas-al-khalidi@utq.edu.iq

Abstract. Metaverse is a high-tech and very advanced tool in the virtual world. Due to the severe lack of research addressing the process of displaying product quality using this tool. This study aimed to examine the importance of individuals using engagement and interaction strategies in the metaverse environment. In order to identify the quality of products within the factory. Taking into account the interaction of other factors, which is the life context that the individual lives in as well as habituation. Data was collected from users of some electronic platforms. The study sample consisted of 199 individuals. The questionnaire was used to collect data and it was also analyzed using the Smart-Plus 0.4, by the PLS-SEM methodology. The most important finding of the study is the effect of Real-time multisensory social interactions (RMSIs) strategies in expressing the reality of the production process in factories and creating the best image of the factory. Taking into account the significance of the interactive variables. Therefore, the study recommended the importance of expanding this model in the future, with the possibility of developing other studies in other geographical locations.

Keywords: Metaverse · RMSIs strategies · Quality Perception of Productivity · Life Context · Habituation

1 Introduction

Many companies today are making unremitting efforts to produce high-quality products (Guo et al., 2023; Alnoor et al., 2022). This requires companies to pay attention to obtaining various resources from the best sources (Kamali et al., 2022). Therefore, companies incur very high costs. The goal of companies today in their success does not end with producing products of the required quality. It needs to build positive perceptions among potential consumers about those products. This is a big problem for companies (Suttikun and Mahasuweerachai, 2023; Atiyah, 2022). The problem stems from the competition that companies experience. Competition requires the company to go to the potential consumer and build positive perceptions about its products (Wang et al., 2022). Many individuals are not convinced simply by seeing an advertisement for a product. Rather, they need information from other parties, such as users (Bayerl and Jacobs, 2023). Some individuals turn to the websites of electronic platforms and search reactions. The

M. Al-Emran et al. (Eds.): IMDC-IST 2024, LNNS 876, pp. 83–93, 2023.
https://doi.org/10.1007/978-3-031-51300-8_6

problem expanded with the introduction of shady and incorrect information. Then, it is possible to research how to address this problem. One of the important digitization tools is the metaverse. Metaverse is a digital tool that simulates reality. Through the metaverse journey, the user can enter into an information-rich experience (Arpaci and Bahari, 2023; Eneizan et al., 2019). Metaverse transports the individual directly into the factory, where the individual learns about the manufacturing stages and sees the smallest details. Unfortunately, there is a clear scarcity of literature that has examined the metaverse. How to use it to build positive perceptions of product quality. Some of these studies investigated the role of metaverses in detecting purchase intentions (jafar et al., 2023; Al-Abrrow et al., 2021). Other studies have investigated the added knowledge to a potential consumer through a metaverse environment (wan et al., 2023; Khaw et al., 2022). Therefore, the current study attempts to shed light on an important aspect of the application of the metaverse in virtual reality. The possibility of research in this field will expand in the future in specific industries or a group of industries of a complex nature. Instead of entering factories and field workshops, it is possible to use the metaverse space. This study added some new insights and ideas into the use of metaverse (Alsalem et al., 2022). By expanding the possibility of its use within the factory corridors. The goal is to build solid perceptions in the consumer's imagination about existing products. Thus, this study made a serious contribution at the intellectual and practical levels in the metaverse environment. In order to address the state of obsession among individuals about the extent of real quality in production factors.

2 Hypothesis Development

2.1 RMSIs Strategies and Quality Perception of Productivity

In the past, the process of interaction between individuals and groups was carried out through personal communications and text messages with solid content (Park et al., 2020). Later, means of communication developed into video calls and meetings conducted through them (Borrelli et al., 2023; Hamid, 2023; Albahri et al., 2021). Especially with the emergence of the Corona pandemic. The process of exchanging ideas and information through these means makes there a greater possibility of attracting the people gathered towards the ideas presented (Anderson et al., 2023; Atiyah and Zaidan, 2022; Fadhil et al., 2021). Which leads to positive perceptions (Alnoor et al., 2022). The Metaverse environment has greatly expanded this attraction and interaction. Metaverse helped build the virtual team. This type of collaborative team usually cannot be created in a physical environment (Lee and Jo, 2023). Due to the impossibility of meeting between team members who live far apart from each other. The discussions and ideas that are raised and exchanged through this type of team contribute better to understanding the quality of productivity of companies working in various specializations (Dwivedi et al., 2022). Immersive experiences, productive content, and digital capabilities have greatly facilitated the perception of quality production. Metaverse reduces geographical, technical, climate and working condition barriers within the factory. Which enables people from all over the world to interact smoothly. This can lead to more in-depth views through live viewing and real-life experience (Wei, 2022; AL-Fatlawey et al., 2021). It can also lead to dealing with people with unlimited experience, with whom dealings are usually not typical, and thus they build better perceptions of the quality. The Metaverse environment encourages the sharing of previous experiences and presenting them in unconventional ways (Pizzolante et al., 2023). This type of presentation makes it possible to visualize the production steps and the possibility of reducing or increasing the quality of the product, and thus there is the possibility of achieving greater perceptions of the topic at hand (Javornik et al., 2021). From the above it can be assumed:

H1: Real-time multisensory social interactions (RMSIs) strategies effect on quality perception of productivity.

2.2 Moderation Role of Life Context

Life context is defined as the conditions that an individual experiences on a daily basis (Barahmand et al., 2019). The context of life means the daily interactions that are necessary for an individual (Granström et al., 2023). These daily conditions and interactions create a mood that reduces or increases an individual's tensions in dealing with various environmental stimuli (Kuehner et al., 2023). An individual's mental health, well-being, physical health, etc., are all factors shaped by the life context in which individuals live (Fotheringham et al., 2023; Atiyah, 2020). When individuals interact with the metaverse environment, they experience immersive and unconventional experiences. These experiences are characterized by unlimited attraction to different compositions in products (Marabelli et al., 2023). This possibility allows the metaverse to create more transparent visualizations. Therefore, it is compatible with individuals who enjoy good mental health. This type of individual's interaction with experiences is unrestrained (Pizzolante et al., 2023). Individuals start asking questions and try to get the best answers in this range. Well-contextualized individuals realize that change in immersive experiences is inevitable. In order to be informed of the best and most sensitive issues for change. Hence better perceptions of product quality (Goldberg and Schär, 2022). The work environment is also a factor that helps shape the context of life. Individuals whose work environment is characterized by noise suffer from the inability to engage in such experiences and feel that it confuses their thinking (Palacios-Ibáñez et al., 2022). On the contrary, individuals who enjoy a calm work environment prefer to deal with such experiences and try to achieve the maximum benefit. The benefit is the possibility of visualizing the quality of the proposed product. Where is it from? How can it be developed better? (Allam et al., 2022). Also, the context of life is expressed in the personal commitments of individuals towards their surroundings. Therefore, the extent to which these individuals can change the reality of their thinking if new ideas about products are presented (Benrimoh et al., 2022). These elements have the potential to moderate the relationship between Real-time multisensory social interactions (RMSIs) strategies and quality perception of productivity. Therefore, we hypothesize:

H2: Life context moderates the relationship between Real-time multisensory social interactions (RMSIs) strategies and quality perception of productivity.

2.3 Moderation Role of Habituation

Habituation means the state that an individual experiences continuously during the various stages of his life (Larson et al., 2023). Habituation is a reprehensible characteristic from many points of view (Becerril-Castrillejo and Muñoz-Gallego, 2022). Habituation is also described as the daily routine that an individual experiences with various social and organizational situations. Many researchers have studied habituation and found that it is a worrying condition that needs to be treated (Houslay et al., 2019). Habituation creates a feeling of lethargy and sluggishness in individuals. When individuals are accustomed to digital interactions, they do not adapt better to the situations at hand (Michailovs et al., 2022; Gatea and Marina, 2016). The metaverse environment elicits new perceptual stimuli. But with habit and lack of change in the methods of presenting topics and ideas, it is not possible to reach correct perceptions about the quality of the product. Some believe that habituation leads to a decrease in sensitivity to external stimuli. It is not strange that this arises when the same routine continues (Leal-Rodríguez et al., 2023; Atiyah, 2020). The time for habituation is also an important element in the natural state of the individual. Some individuals do not adapt easily and are constantly disturbed by external stimuli. As for others, they get used to it for a shorter period and are not reactive to repeated events (Larson et al., 2023; Manhal et al., 2023; Atiyah, 2023). Certainly, the difference in habituation is reflected in the difference in perceptions among individuals regarding the quality of products. Because ideas free from the constraints of habituation are usually characterized by change and are dynamic. Thus, we assume that:

H3: Habituation moderates the relationship between Real-time multisensory social interactions (RMSIs) strategies and quality perception of productivity.

Figure 1 presents the conceptual framework of the present study, to explain the relationship between study's factors. As RMSIs strategies (interpretive variable) effect on quality perception of productivity (responsive variable). As well as, moderative effects of life context and habituation (moderative variables). This framework was constructed based on (Sekaran&Bougie, 2016:75).

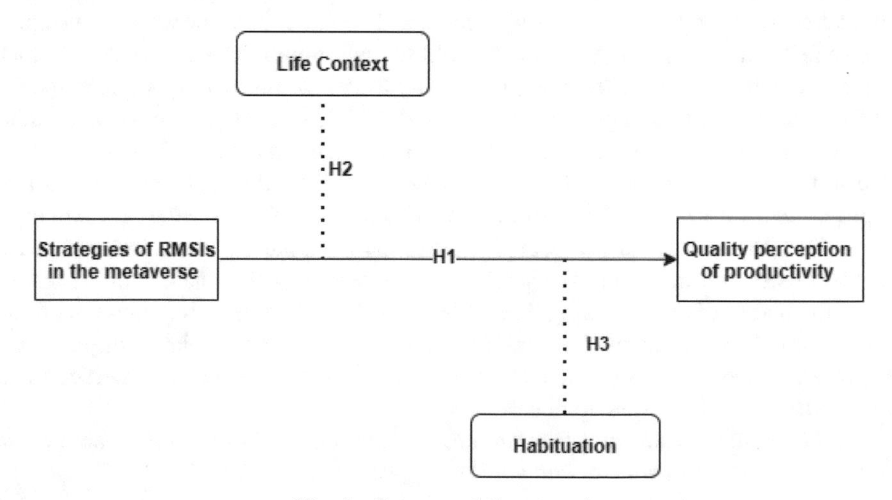

Fig. 1. Conceptual Framework.

3 Research Methodology

In this study, we used a survey approach. As the data was gathered indirectly by the google form questionnaire technique. Because the randomly sample of this study is some of users of social platforms of Roblox. There are some characteristics in data that must be available to meet the requirements of the scale. 1) age of respondents must be more than (35) years, to get clear and accurate answers, 2) the sample must have used social commerce platforms of Roblox for five year or more, to have sufficient experience. The study utilized English scale, thus, translated into Arabic (Brislin's, 1970). To obtain clear items a back-translation tool was utilized by a specialist in both English and Arabic. Responders who completed and resend it were only 218. Questionnaires ineligible were 19. Therefore, the study's data were 199 questionnaires. The participants demographic characteristics are shown in Table 1.

Table 1. Demographics of the participations.

Demographics	Frequency (n = 199)	Percent %
Gender		
Male	*78*	*.39*
Female	*121*	*.61*
Age		
30–39	*81*	*.41*
40–49	*56*	*.28*
50–59	*50*	*.25*
≥ 60	*12*	*.06*
Academic qualifications		
Undergraduate	*114*	*.57*
Post-graduate	*85*	*.43*
Experience		
1–5	*42*	*.21*
5–10	*106*	*.53*
≥ 10	*51*	*.26*

30–39 years represent the highest percentage 0.30. A master's degree or above was held by 0.57percent of respondents, 0.43 percent received an undergraduate degree. 5–10 years represent the highest percentage 0.53 of the experience of the study sample.

The four primary variables in the present study include RMSIs strategies as independent variable, life context and habituation as mediative variables, while quality perception of productivity is the dependent variable. This study adopts well-established scales from the existing literature to measure the variables in the research model. The current research operationalizes RMSIs strategies using 4 items adapted from (Hennig-Thurau et al., 2023) as well as, used 2 items for life context, adopted from (Hennig-Thurau et al., 2023). In contrast, habituation was measured with 2 items adapted from (Hennig-Thurau et al., 2023; Atiyah, 2023). And finally, this study utilized 4 items based on (Castelle, 2017), to measure quality perception of productivity. A 5 point scale (1 = strongly disagree, 5 = strongly agree) was used to operationalize all the concepts.

4 Data Analysis

According to Hair et al. (2010), the measurement model investigates to evaluate the validity and reliability of the created measures. To verify the convergent validity, the average extracted variance and composite reliability should be evaluated. Since they were greater than 0.50, as advised by Hair et al. (2010), the average variance extracted (AVE) values of all latent variables in this investigation ranged from 0.573 to 0.728, which was acceptable. Additionally, as shown by the results in Table 2, the outer loadings of all latent variables for all of the major constructs ranged between 0.913 and 0.719, above the acceptable value of 0.70 as advised by Hair et al. (2017). The latent variables' composite reliabilities also ranged between 0.840 and 0.873. These findings show that the measurement scales used into the model are highly reliable (Hair et al., 2010).

Table 2. Result of measurement model.

Variables	Items	Loading factors	CR	AVE
RMSIs Strategies	M2	0.835	0.840	0.636
	M3	0.803		
	M4	0.753		
Quality perception of productivity	Q1	0.719	0.843	0.573
	Q2	0.726		
	Q3	0.787		
	Q4	0.794		
Life context	L1	0.907	0.873	0.775
	L2	0.853		
Habituation	H1	0.913	0.842	0.728
	H2	0.789		

Table 3 depicts the assessment of discriminant validity using the Fornell and Larcker (1981) criterion. As illustrated, the square root of the AVE of each construct is larger than the correlation estimates of the constructs. This indicates that all the constructs are distinctly different from one another, implying that each construct is unique and captures phenomena not represented by other constructs in the model (Hair et al., 2010).

In this study, three hypotheses were assumed to achieve the study target. For the testing of hypotheses, parameter estimates for statistical significance and coefficient

Table 3. Discriminant validity.

Variables	Habituation	Life context	Quality perception	RMSIs
Habituation	0.853			
Life context	0.699	0.881		
Quality perception	0.382	0.275	0.757	
RMSIs	0.306	0.288	0.403	0.798

values were evaluated using a bootstrapping method based on PLS-SEM (Hair et al., 2017). The bootstrapping method with 5000 bootstrap re-sampling and bias-corrected confidence intervals was used to examine the significance of the path coefficients. Table 4 shows the results of the structural model assessment.

Table 4. Hypotheses test.

Direct path	(O)	(M)	(STDEV)	(O/STDEV|)	P values	R
RMSIs - > Quality perception	0.260	0.273	0.054	4.806	0.000	Yes
Indirect path						
Life context x RMSIs - > Quality perception	0.331	0.119	0.063	5.254	0.024	Yes
Habituation x RMSIs - > Quality perception	0.287	0.074	0.074	3.878	0.043	Yes

The results in Table 4 suggest a direct effect, but there was a strong indication that all statistical assumptions on direct effects were valid. Regarding the moderative influence hypotheses, the results revealed that the moderator variable plays a fully moderating role in the interaction between the variables.

5 Discussion

The results of the study showed the significant role of strategies prepared by companies to enter the metaverse environment. This means that inviting individuals to enter the metaverse should not be random. Good planning and preparation are the basis of success for any distinguished work. This result supports (H1). It coincides with what researchers' opinions have shown about systematic marketing based on preconceived plans and scenarios (Lim, 2023). Many companies launched products and these products quickly disappeared. As a result of not marketing, it well (Fuller, 2008). Thus, companies did not receive full reliability from potential customers. Therefore, it is important to build a good perception of the company's products in terms of quality using metaverse according to strategic plans. The results also showed clear support for (H2). Therefore, we conclude that the life context in which an individual lives interacts positively with both the metaverse and perceptions of product quality. This agrees with the idea that the context of life is the basis for launching the process of interaction with digital tools (Qian et al., 2023). When an individual realizes that the metaverse environment is a spacious space full of abundant information, the individual interacts strongly with the ideas presented. Therefore, can ask questions and get appropriate answers (Lee et al., 2023). Then, can move around the corridors of the places as wants to get to know. On the other hand, the results showed that habituation is an unpleasant element in this relationship. This is support for the (H3). Habituation is one of the factors that contribute to creating a boring routine. The routine does not bring pleasure over time. Thus, companies need to undermine routine in the process of promoting the product. By presenting ideas and information that are modern and exciting. This result is consistent with the study (Kim et al., 2023), which showed that habituation is a negative condition that leads to a reduction in innovation and newer in the company's various operations.

6 Conclusion

From the above we conclude that metaverse strategies play a fundamental role in changing views about the virtual world. Many expect that the metaverse environment is illogical in expressing a good productive system. The metaverse system provides content with tremendous capabilities in representing reality. By depicting real reality in a completely similar format. Companies can attract potential consumers for their products using metaverse capabilities. But there is no marketing for the role of the metaverse space in conveying a live vision of the factory environment in general and production processes in particular. Therefore, this study adopted strategies on how individuals interact and immerse themselves in the metaverse environment according to pre-prepared plans and programs. Thus, it can be said that there is a possibility to adopt these strategies and enhance individuals' engagement with them. There is no doubt that individuals are social entities that influence and are affected by what surrounds them. Additionally, the context of life has a positive interactive role in activating the metaverse effect. When an individual lives within a modern life context, he accepts unconventional tools in expressing the quality of production. And wants to know everything new. The individual feels happy when obtains new information and has a great passion for transmitting information. But at the same time, when an individual gets used to living in a noisy environment with abundant and diverse information, feels bored. Habituation creates a state of routine and lack of innovation in dealing with what is at hand, and thus there is an urgent need for innovation. The metaverse environment is part of a vast digital space that is capable of renewal and innovation. Which helps a lot in crossing the threshold of inactivity in product marketing strategies. By revealing the best image of product quality.

References

Al-Abrrow, H., Fayez, A.S., Abdullah, H., Khaw, K.W., Alnoor, A., Rexhepi, G.: Effect of open-mindedness and humble behavior on innovation: mediator role of learning. Int. J. Emerging Markets (2021)

Albahri, A.S., et al.: Based on the multi-assessment model: towards a new context of combining the artificial neural network and structural equation modelling: a review. Chaos Solitons Fractals **153**, 111445 (2021)

AL-Fatlawey, M.H., Brias, A.K., Atiyah, A.G.: The role of strategic behavior in achievement the organizational excellence" Analytical research of the manager's views of Ur State Company at Thi-Qar Governorate". J. Adm. Econ. **10**(37) (2021)

Allam, Z., Sharifi, A., Bibri, S.E., Jones, D.S., Krogstie, J.: The metaverse as a virtual form of smart cities: opportunities and challenges for environmental, economic, and social sustainability in urban futures. Smart Cities **5**(3), 771–801 (2022)

Alnoor, A., et al.: How positive and negative electronic word of mouth (eWOM) affects customers' intention to use social commerce? A dual-stage multi group-SEM and ANN analysis. Int. J. Hum.–Comput. Interact., 1–30 (2022)

Alsalem, M.A., et al.: Rise of multiattribute decision-making in combating COVID-19: a systematic review of the state-of-the-art literature. Int. J. Intell. Syst. **37**(6), 3514–3624 (2022)

Anderson, C., Baskerville, R., Kaul, M.: Managing compliance with privacy regulations through translation guardrails: a health information exchange case study. Inf. Organ. **33**(1), 100455 (2023)

Arpaci, I., Bahari, M.: Investigating the role of psychological needs in predicting the educational sustainability of Metaverse using a deep learning-based hybrid SEM-ANN technique. Interact. Learn. Environ., 1–13 (2023)

Atiyah, A.G.: Impact of knowledge workers characteristics in promoting organizational creativity: an applied study in a sample of Smart organizations. PalArch's J. Archaeol. Egypt/Egyptol. **17**(6), 16626–16637 (2020)

Atiyah, A.G.: The effect of the dimensions of strategic change on organizational performance level. PalArch's J. Archaeol. Egypt/Egyptol. **17**(8), 1269–1282 (2020)

Atiyah, A.G.: Effect of temporal and spatial myopia on managerial performance. J. La Bisecoman **3**(4), 140–150 (2022)

Atiyah, A.G.: Power distance and strategic decision implementation: exploring the moderative influence of organizational context (2011)

Atiyah, A.G.: Strategic network and psychological contract breach: the mediating effect of role ambiguity (2023)

Atiyah, A.G., Zaidan, R.A.: pp. 115–130). Singapore: Springer Nature Singapore.

Barahmand, N., Nakhoda, M., Fahimnia, F., Nazari, M.: Understanding everyday life information seeking behavior in the context of coping with daily hassles: a grounded theory study of female students. Libr. Inf. Sci. Res. **41**(4), 100980 (2019)

Bayerl, P.S., Jacobs, G.: Who is responsible for customers' privacy? Effects of first versus third party handling of privacy contracts on continuance intentions. Technol. Forecast. Soc. Chang. **185**, 122039 (2022)

Becerril-Castrillejo, I., Muñoz-Gallego, P.A.: Influence of habitual level of consumption on willingness to pay: a satiation, sensitization, and habituation approach. Int. J. Hosp. Manag. **103**, 103210 (2022)

Benrimoh, D., Chheda, F.D., Margolese, H.C.: The best predictor of the future—the metaverse, mental health, and lessons learned from current technologies. JMIR Mental Health **9**(10), e40410 (2022)

Borrelli, S., Fumagalli, S., Colciago, E., Downey, J., Spiby, H., Nespoli, A.: How should a video-call service for early labour be provided? A qualitative study of midwives' perspectives in the United Kingdom and Italy. Women and Birth (2023)

Castelle, K.M.: An investigation into perceived productivity and its influence on the relationship between organizational climate and affective commitment. Old Dominion University (2017)

Dwivedi, Y.K., et al.: Metaverse beyond the hype: multidisciplinary perspectives on emerging challenges, opportunities, and agenda for research, practice and policy. Int. J. Inf. Manage. **66**, 102542 (2022)

Eneizan, B., Mohammed, A.G., Alnoor, A., Alabboodi, A.S., Enaizan, O.: Customer acceptance of mobile marketing in Jordan: an extended UTAUT2 model with trust and risk factors. Int. J. Eng. Bus. Manage. **11**, 1847979019889484 (2019)

Fadhil, S.S., Ismail, R., Alnoor, A.: The influence of soft skills on employability: a case study on technology industry sector in Malaysia. Interdiscip. J. Inf. Knowl. Manag. **16**, 255 (2021)

Fotheringham, P., Dorney, E., McKinn, S., Fox, G.J., Bernays, S.: Protecting mental health in quarantine: exploring lived experiences of healthcare in mandatory COVID-19 quarantine, New South Wales. Australia. SSM-Population Health **21**, 101329 (2023)

Fuller, G.: Consumers, customers, marketplaces and markets in New Product Development–a roadmap through marketing and launching in product development. In: Case Studies in Food Product Development (pp. 354–378). Woodhead Publishing (2008)

Gatea, A.A., Marina, V.: Higher education funding in Iraq in terms of the experience of particular developed countries. Int. J. Adv. Stud. **6**(1), 8–17 (2016)

Goldberg, M., Schär, F.: Metaverse governance: an empirical analysis of voting within decentralized autonomous organizations. J. Bus. Res. **160**, 113764 (2022)

Granström, B., et al.: Perceptions of life and experiences of health care support among individuals one year after head and neck cancer treatment–an interview study. Eur. J. Oncol. Nurs. **66**, 102383 (2023)

Guo, H., et al.: Production of high-quality pyrolysis product by microwave–assisted catalytic pyrolysis of wood waste and application of biochar. Arab. J. Chem. **16**(8), 104961 (2023)

Hamid, R.A., et al.: How smart is e-tourism? A systematic review of smart tourism recommendation system applying data management. Comput. Sci. Rev. **39**, 100337 (2021)

Houslay, T.M., Earley, R.L., Young, A.J., Wilson, A.J.: Habituation and individual variation in the endocrine stress response in the Trinidadian guppy (Poecilia reticulata). Gen. Comp. Endocrinol. **270**, 113–122 (2019)

Jafar, R.M.S., Ahmad, W., Sun, Y.: Unfolding the impacts of metaverse aspects on telepresence, product knowledge, and purchase intentions in the metaverse stores. Technol. Soc. **74**, 102265 (2023)

Javornik, A., et al.: Strategic approaches to augmented reality deployment by luxury brands. J. Bus. Res. **136**, 284–292 (2021)

Kamali, A., Heidari, S., Golzary, A., Tavakoli, O., Wood, D.A.: Optimized catalytic pyrolysis of refinery waste sludge to yield clean high quality oil products. Fuel **328**, 125292 (2022)

Khaw, K.W., et al.: Modelling and evaluating trust in mobile commerce: a hybrid three stage Fuzzy Delphi, structural equation modeling, and neural network approach. Int. J. Hum-Comput. Interact. **38**(16), 1529–1545 (2022)

Kim, N., Grégoire, L., Razavi, M., Yan, N., Ahn, C.R., Anderson, B.A.: Virtual accident curb risk habituation in workers by restoring sensory responses to real-world warning. iScience **26**(1) (2023)

Kuehner, C., et al.: Effects of rumination and mindful self-focus inductions during daily life in patients with remitted depression–an experimental ambulatory assessment study. Behavior Therapy (2023)

Larson, L.M., et al.: Effects of iron supplementation on neural indices of habituation in Bangladeshi children. Am. J. Clin. Nutr. **117**(1), 73–82 (2023)

Leal-Rodríguez, A.L., Sanchís-Pedregosa, C., Moreno-Moreno, A.M., Leal-Millán, A.G.: Digitalization beyond technology: proposing an explanatory and predictive model for digital culture in organizations. J. Innov. Knowl. **8**(3), 100409 (2023)

Lee, H., Yi, Y., Moon, W., Yeo, J.Y.: Exploring the potential use of the metaverse in nurse education through a mock trial. Nurse Educ. Today **131**, 105974 (2023)

Lee, N., Jo, M.: Exploring problem-based learning curricula in the metaverse: the hospitality students' perspective. J. Hosp. Leis. Sport Tour. Educ. **32**, 100427 (2023)

Lim, W.M.: Transformative marketing in the new normal: a novel practice-scholarly integrative review of business-to-business marketing mix challenges, opportunities, and solutions. J. Bus. Res. **160**, 113638 (2023)

Manhal, M., Al-khalidi, A., Hamad, Z.: Strategic network: managerial myopia point of view. Manage. Sci. Lett. **13**(3), 211–218 (2023)

Marabelli, M., Newell, S.: Responsibly strategizing with the metaverse: business implications and DEI opportunities and challenges. J. Strat. Inf. Syst. **32**(2), 101774 (2023)

Michailovs, S., et al.: The impact of digital image configuration on submarine periscope operator workload, situation awareness, meta-awareness and performance. Cogn. Syst. Res. **76**, 13–25 (2022)

Palacios-Ibáñez, A., Navarro-Martínez, R., Blasco-Esteban, J., Contero, M., Camba, J.D.: On the application of extended reality technologies for the evaluation of product characteristics during the initial stages of the product development process. Comput. Ind. **144**, 103780 (2022)

Park, I., Sarnikar, S., Cho, J.: Disentangling the effects of efficacy-facilitating informational support on health resilience in online health communities based on phrase-level text analysis. Inf. Manage. **57**(8), 103372 (2020)

Pizzolante, M., et al.: Awe in the metaverse: designing and validating a novel online virtual-reality awe-inspiring training. Comput. Hum. Behav. **148**, 107876 (2023)

Qian, Z., Day, S.J., Ignatius, J., Dhamotharan, L., Chai, J.: Digital advertising spillover, online-exclusive product launches, and manufacturer-remanufacturer competition. Eur. J. Oper. Res. (2023)

Sekaran, U., Bougie, R.: "Research Methods for Business: A Skill -Building Approach". John Wiley and Sons, Inc. (2016)

Suttikun, C., Mahasuweerachai, P.: The influence of status consumption and social media marketing strategies on consumers' perceptions of green and CSR strategies: how the effects link to emotional attachment to restaurants. J. Hosp. Tour. Manag. **56**, 546–557 (2023)

Wan, X., Zhang, G., Yuan, Y., Chai, S.: How to drive the participation willingness of supply chain members in metaverse technology adoption? Appl. Soft Comput. **145**, 110611 (2023)

Wang, E., Gao, Z., Heng, Y.: Explore Chinese consumers' safety perception of agricultural products using a non-price choice experiment. Food Control **140**, 109121 (2022)

Wei, D.: Gemiverse: the blockchain-based professional certification and tourism platform with its own ecosystem in the metaverse. Int. J. Geoheritage Parks **10**(2), 322–336 (2022)

Factors Influencing School Teachers' Intention to Adopt Open Virtual Educational Resources Platform in Saudi Arabia

Waleed Saud Alshammri, Siti Mastura Baharudin(✉), and Azidah Bt Abu Ziden

School of Educational Studies, Universiti Sains Malaysia, Penang, Malaysia
waleedalshammari@student.usm.my, {sitimastura,azidah}@usm.my

Abstract. The main goal of this study was to look at the factors that influence instructors' intentions to use open educational resources (OER). According to the theoretical framework, Saudi instructors' intentions to utilize OER will be significantly influenced by internal factors like performance expectancy and effort expectancy as well as external elements like social influence and enabling environment. The survey method was used to gather the data, and a two-step structural equation modeling method was used for analysis. It was discovered that performance expectations, effort expectations, and social influence were the factors that most strongly influenced schoolteachers' intentions to utilize open educational resources (OER). There were 360 valid replies in total. This study offers a solid theoretical framework for analyzing teacher behavior from a scholarly standpoint. Additionally, the viewpoints of the schoolteachers provide all educators with useful information for improving OER platforms.

Keywords: OER platform · intention to adopt · schoolteachers · UTAUT · Saudi Arabia

1 Introduction

Open Educational Resources (OER) encompass educational, research, and instructional materials, regardless of their format, that are publicly available and have been made accessible to others without significant restrictions or constraints (Kurelovic, 2020). OER brings a shift in educational institutions allowing equity in access to educational resources worldwide, access without barriers, and flexibility in the realm of both teaching and learning (Kumar et al., 2019; Tang, Lin, & Qian, 2020; Al-Abrrow et al., 2021). In light of the implementation of lockdown measures during the Covid-19 outbreak, educational institutions have transitioned to online learning and instruction. As a result, researchers have expressed a keen inclination towards the utilization of Open Educational Resources OER (Menzli et al., 2022; Albahri et al., 2021). SHMS, an OER platform that denotes "sun" in Arabic, was established in Saudi Arabia as a reliable and secure OER tool that provides Arabic academic resources for students, faculty, and teachers as part of the learning process (SHMS, 2019). The Ministry of Education has mandated that all Saudi Arabian school teachers must officially enroll and use the OER SHMS platform

M. Al-Emran et al. (Eds.): IMDC-IST 2024, LNNS 876, pp. 94–111, 2023.
https://doi.org/10.1007/978-3-031-51300-8_7

as a crucial part of their pedagogical preparation and academic endeavors (SHMS, 2019; AL-Fatlawey et al., 2021).

In this situation, the majority of teachers have emphasized the significance of efficient advertising and promotion as well as the supply of numerous motivating factors to encourage their use of the OER SHMS platform (Al-Shamrani, 2019; Alnoor et al., 2022). Some teachers have voiced worry that introducing OER into their curriculum will cause practical problems. Open educational resources (OER), in the opinion of some experts, have a greater impact on teachers' educational experiences when they have a positive intention than when they exhibit acceptable behavior (Tang et al., 2021; Kurelovic, 2020; Alsalem et al., 2022). The likelihood that teachers will use OER in their future instruction depends on their goals for doing so (Cai et al., 2023). Consequently, a gap in research exists that need addressing in order to facilitate the integration of technology into classroom settings by educators (Antonietti et al., 2022). The Unified Theory of Acceptance and Use of Technology (UTAUT) framework elucidates the inclination of individuals to utilize Open Educational Resources (OER) technology within the realm of the educational process (Al-Mamary, 2022; Antonietti et al., 2022; Kurelovic, 2020; Atiyah, 2020). By focusing on the internal and external factors for adoption intentions, the UTAUT model (Venkatesh et al., 2003) can be used as a potential lens to investigate the essential variables influencing teachers to adopt OER at the individual level (Padhi, 2018; Tang et al., 2020; Villanueva & Dolom, 2018). Performance and effort expectations, among other things, are crucial factors that influence people's willingness to use OER in Saudi Arabia (Al-Shamrani, 2019), which includes Venkatesh et al. (2003), in voluntary contexts. These components, according to earlier researchers, should be examined in the context of Saudi Arabia. The original version of the UTAUT model was employed to validate the substantial positive impact of performance and effort expectancy on the desire to utilize Open Educational Resources (OER), without taking into account any moderator factors. Among other things, social influence should be considered in order to broaden teachers' behavioral intention to use OER (Doan & Dao, 2020; Kurelovic, 2020). According to Shamrani (2019) and Smirani and Boulahia (2022), facilitating conditions among teachers appear to be an important factor that influences users' acceptance of OER use and should be investigated in the Saudi context.

However, only a few studies have comprehensively looked into these behavioral intention determinants in the context of OER adoption. In order to better understand teachers' intentions to accept OER, research is also necessary for other crucial aspects (Kurelovic, 2020; Padhi, 2018), particularly when viewed from Saudi Arabia's perspective (Al-Shamrani, 2019; Manhal et al., 2023). More specifically, it is suggested by Al-Shamrani (2019) and Jung and Lee (2020) that social influence may play a significant role in teachers' intention to use open educational resources (OER), despite mixed research results. Further stressing the significance of the examination of effort expectancy as one of the teachers' positive affective responses to adopting OER is Tseng et al.'s statement from 2022. While the majority of recent studies have concentrated on analyzing the intention to use open educational resources (OER) as the main research outcome (Kurelovic, 2020; Tang et al., 2021), little attention has been given to investigating the connection between this intention and the four factors of the UTAUT in a single model within the context of OER in education. Additionally, prior studies that used UTAUT had limitations due to their scant examination of teachers in schools, specifically those in universities (Doan & Dao, 2020; Kurelovic, 2020; Khaw et al., 2022). By investigating the acceptance of OER in Saudi Arabia, including school teachers, this proposed study filled in the gaps. By examining the impact of these factors on school teachers' intentions to use OER, the study aims to close the gap. In this context, the study aims to pinpoint the internal and external elements that significantly affect the acceptance of the OER platform, offering all educators useful information for enhancing and maintaining OER-related platforms. By utilizing the UTAUT model, the study aims to investigate the effects of the OER SHMS platform factors on the teachers' intention to embrace its use.The present article is structured into distinct sections. The next section begins with the literature review which presents an overview of recent advancements in OER platform research. Four significant factors are discussed theoretically, and hypothetical relationships are established with teachers' intentions to adopt the OER platform. The following section presents the study's methods, and results, and interprets the findings. Finally, the implications are deliberated and conclusively summarized.

2 Literature Review

OER platforms were introduced by UNESCO in 2002, enabling practice groups, such as students, teachers, and learners, to freely copy, modify, and share their resources (Alyami, 2020; Smirani & Boulahia, 2022). Open educational resources (OER), according to De Hart et al. (2015), are books that are either freely accessible to the general public or have a Creative Commons (CC) license. According to Kim (2008), these licenses enable the sharing and modification of Open Educational Resources (OER) and facilitate collaboration for the purpose of reuse, particularly in commercial contexts. The global Open Educational Resources (OER) movement has had a notable surge in momentum in recent times, mostly driven by the growing recognition and use of online learning and education. This has led to the emergence and implementation of various OER projects, as evidenced by the works of Kumar et al. (2019) and Tang et al. (2021). The adoption of the Open Educational Resources (OER) movement in Saudi Arabia was undertaken by the Center for E-Learning and the OER Network (SHMS) on March 10, 2017 (National Center for E-Learning, 2017). It provides 52,788 courses, 24,929 learning activities, and 378,523 free educational resources (Tlili et al., 2021). For those who seek it, SHMS offers free, open, and cutting-edge learning opportunities as well as fresh insights into integrating technological tools into the educational process (Al-Hchaimi et al., 2022; Chew et al., 2023; Abbas et al., 2023).

2.1 UTAUT Model

The UTAUT model is used in this study to clarify and analyze the teachers' intentions to adopt OER. One of the most thorough theories that can aid in deriving predictions and explanations for the adoption of technology is UTAUT. The UTAUT model uses four central integrated variables to analyze users intentions and effort expectancy (Alasmari, 2017; Tarhini et al., 2015). According to Daniali et al. (2022), the main goal of UTAUT is to make users' intentions regarding the adoption and sustained use of a new information system clear. Additionally, numerous studies have employed UTAUT to comprehend and forecast the variables that affect users' intention to use technology on a global scale (Arif et al., 2018; Hamid et al., 2021). However, in order to strengthen the UTAUT's applicability in the context of school teachers, there must be additional effective factors on their end. A longitudinal study was carried out by Venkatesh and colleagues (2003) to compare eight models and theories empirically among employees who had been exposed to new technologies in four different organizations. However, it was found that three constructs (namely, self-efficacy, anxiety, and attitude) did not directly influence technology adoption after analyzing the comparison's final results. Following the elimination of three constructs, the initial analysis identified seven constructs as significant direct predictors of technology acceptance and usage. The authors of the study meticulously considered the four essential components of the final UTAUT model, namely performance expectancy, effort expectancy, social influence, and facilitating factors. The UTAUT methodology is utilized in this study to assess the impact of the four major ideas on instructors' inclination to utilize the OER SHMS platform.

2.2 Hypotheses Development

According to the UTAUT model, performance expectancy refers to how much a person anticipates that using a system will improve their ability to perform at work (Venkatesh et al., 2003). According to Al-Abdullatif and Alsubaie in 2020 and Rahmaningty et al. in 2020, performance expectancy plays a significant role in how individuals behave when it comes to accepting digital technology in educational settings (Doan & Dao, 2020; Mohan et al., 2020; Padhi, 2018; Gatea and Marina, 2016). Therefore, if they feel that OER offers them a wide range of crucial advantages, teachers are more likely to adopt it. In light of this, the following hypothesis is posited:

H1: Performance expectancy positively impacts the behavioural intention to use the OER.

An intrinsic motivator is effort expectancy, which is the process of making it possible to achieve reputable results when using new technology, like OER. The degree to which people view technology as user-friendly is thought to be related to behavioral intentions regarding adoption and use of the technology (Al-Abdullatif & Alsubaie, 2020). Additionally, a number of researchers have found a positive correlation between effort expectancy and the intention to use OER (Kurelovic, 2020; Li & Min, 2021; Doan & Dao, 2020; Fadhil et al., 2021). The amount of work required when teachers use an OER to find resources and use it will have an impact on their decision to adopt the OER platform. Numerous studies conducted in the Saudi Arabian context have also demonstrated the significant impact of effort expectancy on behavioral intentions toward the use of technology.

H2: Effort expectancy positively impacts the behavioural intention to use the OER.

A teacher's perception of support from their principal, school, friends, and other supervisors who have a significant say in the use of open educational resources (OER) for teaching and learning is referred to as their social influence. Potential adopters may experience uncertainty with regard to the anticipated outcomes of implementing the OER platform and, as a result, may seek the counsel of their influential social connections before making an adoption decision. Social influence has been taken into account in earlier studies on innovation in technology and has been acknowledged as a key driver of behavior in the adoption of Web 2.0 technologies (Teo et al., 2019; Atiyah and Zaidan, 2022). Al-Mamary (2022) asserts that teachers' intentions to use learning management systems (LMS) platforms will be more influenced by other people's opinions. Social influence significantly affects behavior intention to use MOOCs in the context of OER, according to Chaveesuk et al. (2022). Based on this literature review, the following hypothesis is proposed:

H3: Social influence positively impacts the behavioural intention to use the OER.

According to Venkatesh et al. (2003), the enabling condition is the extent to which an individual holds a sincere belief in the progress of a well-structured, technologically advanced, and economically viable infrastructure that facilitates the adoption of a novel system. Hence, this study defines the utilization of the Open Educational Resources (OER) SHMS platform for instructional purposes by Saudi educators as a means of enabling conducive circumstances, wherein they take into account the accessibility and availability of educational resources and technological infrastructure. In particular, facilitating conditions, which are external factors, have been shown to be effective predictors of the OER adoption process. These conditions include giving schools access to the resources they need to access OER, technical support, and unrestricted internet access (Doan & Dao, 2020). Facilitating conditions have a significant influence on teachers' behavioral intentions to use MOOCs in addition to playing a significant role in innovation adoption (Tseng et al., 2022; Atiyah, 2023). The behavioral intentions of teachers to use MOOCs are also significantly impacted (Tseng et al., 2022). Adoption of innovation depends on favorable conditions. In light of this, it can be said that the behavioral intention to adopt OER increases with higher facilitating conditions. The following hypothesis is proposed:

H4: Facilitating conditions positively impact the behavioural intention to use the OER.

2.3 Conceptual Framework

The UTAUT model, which provides a better understanding of behavioral intention to use OER, is adopted in this study. The approach that emerged in the literature to study the acceptance of technologies served as the foundation for the theoretical framework created to represent the relationships of the UTAUT variables (Alasmari, 2017; Tarhini et al., 2015). Therefore, the conceptual model includes five constructs as independent variables for the teacher's behavior intention toward OER adoption: performance expectancy, effort expectancy, social influence, and facilitating conditions. The current study, which is grounded in a critical review of the literature, aims to determine the influence of social influence, supportive environments, as well as internal and external factors, like effort expectations and performance expectations, on Saudi teachers' intentions to use open educational resources (OER).

3 Methodology

OER SHMS platform is a newly established online community providing access to quality teaching and learning resources openly licensed for students and instructors. This quantitative, explanatory study uses empirical data to test the UTAUT theory in order to explain and predict the expected relationship between internal and external factors and an observed dependent variable (intention to adopt OER). The framework and hypotheses were developed using discussions from the literature review. The study then created a questionnaire to collect the data required for the analysis of the hypotheses put forth as well as for testing and validating the ultimate framework for the intention to adopt OER in Saudi Arabia. Given the purpose of the study, Saudi school teachers who actively use the OER SHMS platform and have at least some experience using platforms for educational purposes make up the target population of interest.

Saudi teachers who actively use the OER SHMS platform are given a questionnaire to complete online. The survey is distributed using Google Forms and social networking sites (SNSs, such as WhatsApp, etc.), along with a cover letter outlining the purpose of the study and the survey's guidelines. The online survey was determined to be the most appropriate method because it provided access to participants from various Saudi Arabian schools without significantly increasing costs, risks of infection, or time, money, or both. To gather empirical data, an online survey was conducted utilizing a 22-item questionnaire (refer to Table 1 in the Appendix). The authors conducted this study over three weeks, completing the data collection process on November 2021, resulting in a total of 400 questionnaires being collected. Upon analysis, it was discovered that 40 participants lacked experience in online teaching, rendering them unsuitable for inclusion in this study and thus were excluded. After 40 questionnaires were eliminated, 360 were found to be valid and were used as the final dataset for the analysis that followed. 54.7% and 45.3% of the sample were male and female teachers, respectively. According to the findings, 132 respondents (36.7%) have used the OER SHMS platform for two years, 102 (28.3%) for three years, and 20.6% for one year.

Three components make up the questionnaire for this study, which adapts closed-ended or structured questions. The questionnaire was created and designed using data from prior research. For each construct, researchers assigned measurement items. The first variable, the performance expectancy, has five items: OER SHMS platform useful in sharing learning materials, accomplishing teaching tasks more quickly, improving the quality of teaching tasks, increasing productivity in teaching, and increasing my chances of getting a better position (Alasmari, 2017; Nandwani & Khan 2016; Venkatesh et al., 2003).

The second construct, effort expectancy, used five items, including ease of understanding, ease of becoming skilled at using SHMS, and ease of learning how to operate the OER SHMS platform (Alasmari, 2017; Nandwani & Khan 2016; Venkatesh et al., 2003). In addition, four items—including "people's influence to use the OER SHMS platform" and "managers in school have helped to use SHMS"—were used to gauge social influence (Alasmari, 2017; Nandwani & Khan 2016; Venkatesh et al., 2003). The fourth variable, facilitating conditions, is measured by four factors: the tools required to use the OER SHMS platform, the user's knowledge, the system's compatibility with other OER systems, and the availability of help (Alasmari, 2017; Nandwani & Khan,

2016; Venkatesh et al., 2003). The last variable, behavioral intention, is measured by three items that were taken from Venkatesh et al. (2003) and which include the intention, prediction, and planning to use the OER SHMS platform.

The study followed a particular protocol in order to maintain the validity of the online questionnaire. Three academic experts in education technology were given access to the study's adopted instrument's original Arabic draft at the start of the investigation, along with a brief explanation of the study's history and its objectives for item validation. To identify questions that were confusing or difficult to comprehend, a pilot study using 30 respondents and the personal interviewing approach was also carried out. The final data-gathering process used the modified questionnaire once it was determined that the results were reliable based on Cronbach's alpha.

4 Data Analysis

According to Hair et al. (2019), the partial least squares structural equation modeling (PLS-SEM) method was found to be more suitable for estimating causal relationships among various independent and dependent constructs. The SEM-PLS approach, which is based on a number of steps, is the data analysis technique that was used. First, the measurement model is evaluated, which considers indicators of indicator reliability (such as outer model loading and composite reliability (CR)), discriminant validity, and convergent validity. The second step includes evaluating the analysis of the structural model performed using the bootstrapping method with 5000 bootstrap resampling. The structure model has four components: R2, Q2, the inner model path coefficient, the effect size (f2), and the endogenous variable variance (R2).

5 Results

Discriminant validity and convergent validity are two ways to assess construct validity (Hair et al., 2019). According to Hair et al. (2019), Table 1 and Fig. 1 show that each item's factor loadings should be 0.708. However, if the AVE and CR are met, researchers may take into account alternative item loadings between 0.40 and 0.70. All of the outer model loadings in our study, which ranged from 0.439 to 0.955, were within the acceptable range. Cronbach's alpha coefficients and the CR were used to assess the construct reliability. Additionally, the CR value should be at least 0.7 and the AVE value should be at least 0.5, according to Hair et al. (2019). Our AVE values ranged from 0.510 to 0.836, while our CR values ranged from 0.851 to 0.939, both of which were higher than the 0.7 cutoff.

The Heterotrait-Monotrait Ratio (HTMT) was employed to establish the discriminant validity. Henseler et al. (2015) have established that discriminant validity can be established by ensuring that the HTMT value for each construct is below the cautious threshold value of 0.85 percent. The maximum value of a construct found was 0.843, demonstrating discriminant validity, as shown in Table 2's HTMT values for five constructs, all of which fall below the cutoff of 0.850. In this study, it is clear that one latent variable has already taken into account the variability of its corresponding indicators,

Table 1. Measurement model.

Construct	Item	Loading	CA	CR	AVE
Behavioural intentions (BI)			0.901	0.939	0.836
	BI1	0.867			
	BI2	0.955			
	BI3	0.918			
Effort expectancy (EE)			0.856	0.891	0.579
	EE1	0.598			
	EE2	0.792			
	EE3	0.810			
	EE4	0.722			
	EE5	0.822			
	EE6	0.797			
Facilitating conditions (FC)			0.702	0.815	0.528
	FC1	0.621			
	FC2	0.712			
	FC3	0.863			
	FC4	0.689			
Performance expectancy (PE)			0.745	0.833	0.51
	PE1	0.674			
	PE2	0.823			
	PE3	0.783			
	PE4	0.781			
	PE5	0.439			
Social influence (SI)			0.765	0.848	0.583
	SI1	0.694			
	SI2	0.801			
	SI3	0.801			
	SI4	0.754			

outperforming the variability of other latent variables. As a result, the measurement model has successfully undergone validation based on the validity and reliability tests.

The structural model was computed using the Bootstrap method, employing 5000 samples and a significance threshold of 5%, for the purpose of conducting a hypothesis test. This study utilized the inner variance inflation factor (VIF) values to compute collinearity, which is one of the criteria (Hair et al., 2019). However, the optimal VIF values should be between 3 and 3, as collinearity problems can also occur when two

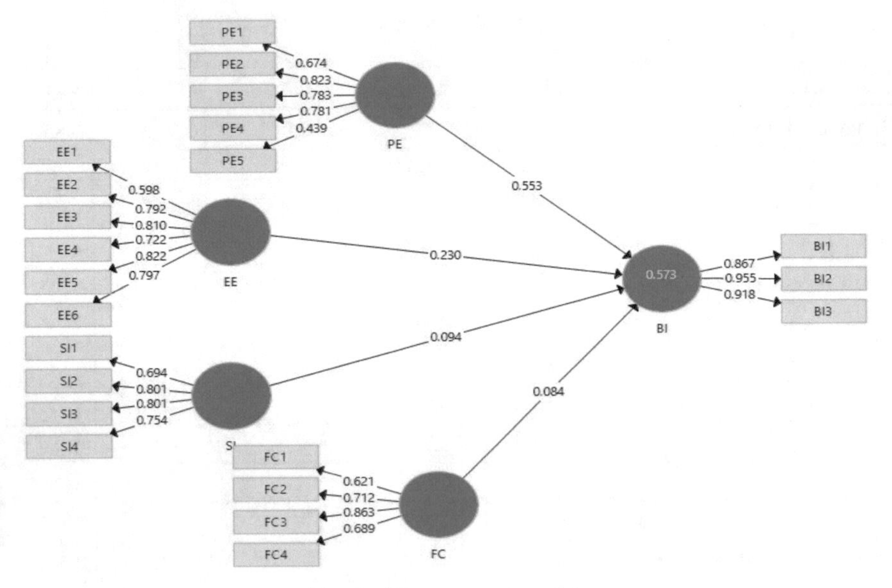

Fig. 1. Measurement model.

Table 2. Discriminant Validity.

	BI	EE	FC	PE	SI
Behavioural intentions (BI)					
Effort expectancy (EE)	0.507				
Facilitating conditions (FC)	0.625	0.793			
Performance expectancy (PE)	0.843	0.377	0.634		
Social influence (SI)	0.422	0.266	0.363	0.490	

variables are highly correlated and a VIF of 3–5 (Hair et al., 2019; Zaidan, & Arianpoor, 2022). According to Table 4, the "Inner VIF Values" of the exogenous latent constructs varied between 1.166 and 1.921, all of which were below the threshold of 3.0. Therefore, it can be inferred that the presence of collinearity did not pose a problem in the present investigation.

As a result, the examined study presents a structural model (refer to Fig. 1) which indicates that social influence, performance expectancy, facilitating conditions and effort expectancy collectively account for 57% of the variability observed in the behavioral intention to utilize open educational resources (OER) (R2 = 0.573). This leads to the conclusion that, in accordance with the recommendation, the endogenous latent constructs had a sufficient level of R2 values. The findings demonstrate that the model, with a Q2 value of 0.456, had a strong predictive relevance. Given that the values of the Q2 were greater than 0, it can be concluded that the exogenous constructs have predictive value for the endogenous constructs, including behavioral intention to use OER.

Table 3 and Fig. 2 display the structural coefficient estimates between the latent variables for each model. Hair et al. (2019) suggested using path analysis to evaluate the relationships between the latent factors. Three of the four hypotheses were confirmed. The outcomes of the structural model are shown in Table 3, along with the standardized path coefficients and corresponding t-values for the paths. First, hypotheses 1 to 2 were confirmed in terms of how internal factors affected teachers' intentions to adopt OER. The decision to utilize the Open Educational Resources (OER) platform was influenced by two significant variables: performance expectancy ($\beta = 0.553$, $p < 0.01$) and effort expectancy ($\beta = 0.230$, $p < 0.01$). The results of this investigation offer empirical evidence in favor of hypotheses H1 and H2. Hypothesis 3 (H3) was substantiated by the finding that social influence had a substantial impact on instructors' intentions to utilize the Open Educational Resources (OER) platform, as evidenced by the external factors ($\beta = 0.094$, $p < 0.01$). There was no statistically significant variation in the intents to utilize the Open Educational Resources (OER) platform when considering the different facilitation settings ($= 0.084$, $p > .05$). The lack of support for hypothesis H4 resulted in its rejection. The findings of the study demonstrate that a majority of the hypotheses were corroborated, and the theoretical framework proposed in the research elucidated 57% of the variance in the participants' behavioral inclination to utilize Open Educational Resources (OER). This discovery suggests that the study offers a satisfactory theoretical justification for the adoption model of Open Educational Resources (OER) platforms among educators. The outcome of the structural model is illustrated in Fig. 2.

Table 3. Structural Model.

	Statement	Standard beta	t value	P
H1	Performance expectancy - > behavioural intentions	0.553	14.57**	0.000
H2	Effort expectancy - > behavioural intentions	0.230	5.452**	0.000
H3	Social influence - > behavioural intentions	0.094	2.597**	0.009
H4	Facilitating conditions - > behavioural intentions	0.084	1.632	0.103

* p < .05; ** p < .01

In addition to looking at the R2 values of all the endogenous constructs, f2 can be used to gauge the extent to which an exogenous construct has an effect on the endogenous constructs. Hair et al. (2017) delineated distinct classifications of effect size, with those measuring at 0.35 and above being deemed large, those from 0.15 to 0.34 medium, and those up to 0.14 having a small effect size. The results about the effect size are outlined in Table 4. Based on the outcomes, performance expectancy was identified as the model's more significant factor influencing behavioral intentions. Concurrently, effort expectancy and social influence were observed to have a near-to-small effect. This demonstrates the current study's high relevance and strong influence on the dependent

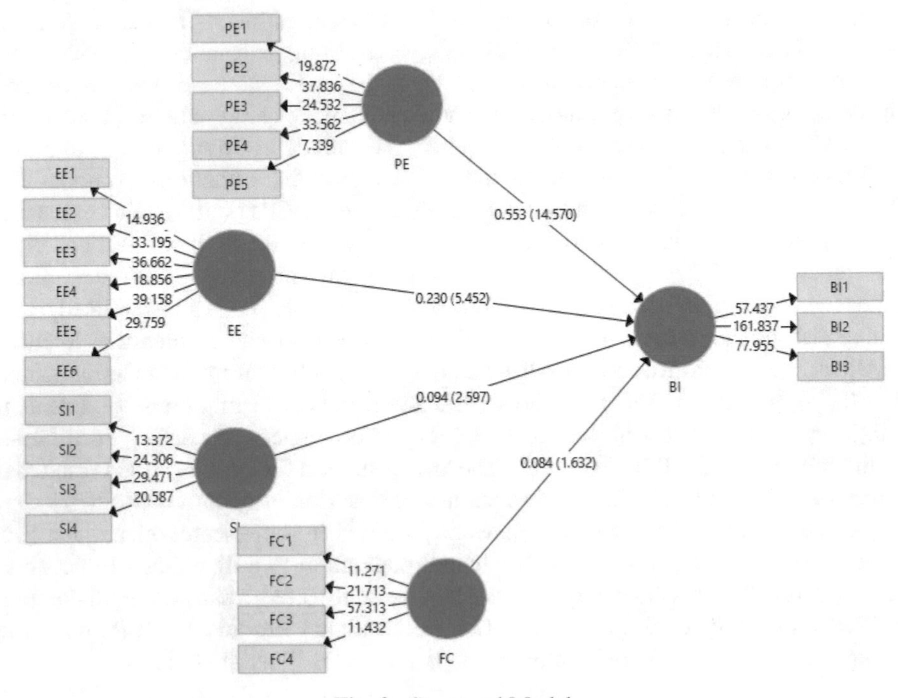

Fig. 2. Structural Model.

variable. These findings highlight the greater factors that impact behavioral intentions to use OER.

Table 4. Effect Size.

	Behavioral intention	Effect
Performance expectancy	0.504	Large
Effort expectancy	0.076	Small
Social influence	0.018	Small
Facilitating conditions	0.009	Small

6 Discussion

This study investigates the link between teachers' intention to use an OER platform and their performance expectations. According to the relationship proposed and significance values shown in Table 3, the path was significant and positive, supporting the theoretical hypothesis (H1). This result demonstrates that if the OER SHMS platform is profitable, efficient, effective, and productive, teachers will use it. This is in line with the majority of earlier studies that forecast the connection between behavioral intention and performance expectancy to implement the technology (Doan & Dao, 2020; Mohan et al., 2020; Padh, 2018; Venkatesh et al., 2003). The findings show that Saudi school teachers are satisfied with the current online resources and are aware of the advantages of the OER SHMS platform in enhancing learning and teaching. Saudi teachers use the OER platform for instruction, as this study's findings demonstrate, because they think doing so will help them perform better and learn more quickly. This result further corroborates Al-Shamrani (2019)'s assertion that efficiency and effectiveness are the primary motivators for OER SHMS platform adoption among all Saudi teachers. The strongest predictor of Saudi teachers' intention to use OER, with the largest effect size on the dependent variable, is performance expectancy. This implies that the advantages of the OER SHMS platform are the main factors driving the teacher's decision to use the platform for instruction.

It was discovered that there was a significant and favorable relationship between teachers' expectation of effort and their intention to use OER, supporting hypothesis H2. Teachers are willing to use the OER platform because they believe it will be simple to use, according to numerous previous studies (e.g., Alghamdi et al., 2022; Al-Mamary, 2022; Kurelovic, 2020; Li & Min, 2021). Given that the school teachers used the platform for their teaching, this may help to explain the significant influence of effort expectations. The construct may therefore be relevant at this time, especially for Saudi teachers who are already familiar with how OER can be used for instructional activities (Al-Shamrani, 2019). This demonstrates that they will have access to a variety of useful and trustworthy resources if using the OER SHMS platform is convenient and simple. It is important to note that Saudi teachers prioritize their own efforts by promoting and encouraging the use of the OER platform, as well as by improving its appearance and functionality. This has the effect of enhancing teachers' perceptions of OER's usability. Teachers might think that the OER SHMS platform is very approachable and user-friendly and that this will improve the caliber of their instruction.

The study's findings indicate a favorable and significant correlation between social influence and teachers' intention to use open educational resources (H3). This supports the assertions made by Venkatesh et al. (2003) about the importance of social influence in technology adoption. It is consistent with the earlier hypothesis that the strength of social networks will increase teachers' intentions to use OERP in the future (Al-Mamary, 2022). Studies conducted in educational settings have demonstrated that teachers' social networks, which include their school administrators and coworkers, have a significant impact on their intentions to use technology to support their students' learning (Chaveesuk et al., 2022; Teo et al., 2019). The outcome confirms the findings of earlier studies, which showed that teachers' intentions to use the OER SHMS platform are attributed to advice and recommendations from their colleagues and school administrators who think the platform should be used in the classroom. Additionally, school administrators have been crucial in helping teachers use the OER platform in their lessons. The managers may encourage the teachers to spend more time learning from the OER platform, which makes this scenario plausible. As a result, teachers are more likely to adopt the OER platform the more important the managers' recommendations are to them. Accordingly, it can be deduced that social influence is one of the main factors Saudi teachers take into account when deciding whether to use the OER SHMS platform in their classrooms (Alkhasawneh, 2020).

This finding is inconsistent with prior research suggesting that facilitating conditions directly impact teachers' intentions to adopt mobile technology (Doan & Dao, 2020; Tseng et al., 2022). The reason is that facilitating conditions generally influence the actual usage of OER rather than behavioral intention (Kurelovic, 2020). A UTAUT model, showed that facilitating conditions relate to actual usage more than behavioral intention (Venkatesh et al., 2003). This finding addresses the concern that Kurelovic (2020) raised, who explained that facilitating conditions are related to external factors and better predict the actual use of OER than behavioral intention. To some extent, the insignificant impact of facilitating conditions on intention may be a methodological artifact: teachers are less likely to capture facilitating conditions at the moment of the survey. Thus, it is unsurprising that facilitating conditions have emerged as a decisive factor in teachers' willingness to adopt OER.

6.1 Theoretical Implications

Despite the extensive body of literature on teachers' intention to adopt OER, there exists a dearth of information about the interrelationships between the variables of the UTAUT as a fundamental theoretical framework. Additionally, only a handful of studies have explored the role of the four fundamental components of UTAUT and their impact on the intention to adopt OER (e.g., Kurelovic, 2020; Tang et al., 2021). Thus, the primary theoretical contribution of this investigation is to scrutinize these factors in the context of OER adoption intention by incorporating the UTAUT model in a non-Western country, such as Saudi Arabia, which has limited experience with OER adoption technology, specifically in the educational context of school teachers. This finding supports earlier studies that emphasized performance expectations, effort expectations, and social influence as the main factors influencing OER acceptance. Combining predictor variables from the suggested research model can account for up to 57% of the variance in intention to use OER. The proposed theoretical framework created to examine teachers' intentions to adopt OER using UTAUT is where this study gets its theoretical contribution. The results showed that the research model has good explanatory power in predicting Saudi teachers' intention to adopt OER.

6.2 Practical Implications

The Saudi Ministry of Education has the ambition to stimulate technology integration across all levels of instruction, in line with the Saudi Vision 2030 plan focused on revolutionizing education. The Saudi government fully supports the improvement of education by offering top-notch learning materials like the OER SHMS platform. However, before the start of this study, it was clear that little was known about the factors impacting teachers' inclination to use OER in the Saudi context (Alkhasawneh, 2020). The results of this study will therefore benefit the government by offering solid justification for the security guarantee in OER adoption. This study offers a model that could be used by the government and the OER SHMS platform to better understand the reasons why schoolteachers choose to use the platform and the factors that influence those choices. Designers of the OER SHMS platform repository can benefit from this observation by improving the user experience to increase teachers' effectiveness and productivity. When making strategic decisions to support teachers' perceived ease of using OER, repository designers must take performance expectancy and effort expectancy into account. Teachers exhibit a preference for conveniently obtainable resources. Therefore, streamlining the exploration procedure, through the automatic filtration of repeated data and the suggestion of resources that are suitable for a particular age, grade, or subject, could prove beneficial (Tang et al., 2020). For administrators, it is recommended that government or Ministry of Education can integrate OER training in teacher education training programs to enhance the creation/adapting and redistribution of OER so that their performance expectancy and effort expectancy can be strengthened. One potential method for increasing OER platform awareness among teachers is through advertising on social media platforms such as Facebook, Snapchat, Instagram, and Twitter.

In light of the limited impact of facilitating conditions on the intention to utilize the OER platform, we propose that technological resources and individuals (or experts) within educational institutions offer instruction and support to teachers through the OER platform. This approach may engender distinctive levels of platform adoption and enhancements in specific features. Both of these practices can enhance teachers' willingness to adopt the OER platform or its advanced version and improve intentions with the OER platform.

It should be noted that this study was conducted solely from the school teachers' perspective. However, it is important to consider that students also have the potential to utilize OER, especially as many students exclusively rely on mobile devices for searching and studying. Hence, further research could be conducted to explore the effectiveness of OER from students' perspectives.

7 Limitations and Future Research

The current study has several limitations that need further exploration. Firstly, to obtain primary data, the study solely relied on conducting surveys among school teachers in Saudi Arabia. Secondly, the current study was conducted within the SHMS platform, an innovative OER platform that aims to improve educational material and promote education in Saudi Arabia. While this platform is a significant development, it may be beneficial to test the generalizability of the findings by incorporating other LMSs into the study. Moreover, another important limitation concerning the research the antecedents of intention to adopt OER were limited to four internal and external factors of the UTAUT model. Although the R2 values are considered highly acceptable by behavioral research standards (Hair et al. 2019), there may be other factors influencing teachers' intention to use OERP beyond those investigated in the present study. To increase the explanatory power and generalizability of the model, future studies should think about incorporating additional variables, such as anxiety and self-efficacy. It is advised that more research be done using an experimental design to compare the different effects on both male and female populations regarding the UTAUT model. This is because different genders have different effects on people's intentions to adopt.

8 Conclusion

The present study aimed to fill a gap in the existing literature by investigating the many internal elements (namely, effort expectations and performance expectations) and external factors (namely, social influence and resource facilitation) that impact the intention of school instructors to utilize open educational resources (OER). The criteria can be obtained by employing the Unified Theory of Acceptance and Use of Technology (UTAUT). The suggested model demonstrated significant explanatory capacity through empirical testing. The outcomes of this study may be of utility to educators in Saudi Arabian schools as well as administrators of public and open-access SHMS platforms. Based on the results of the study, it was observed that Saudi educators often employ the Open Educational Resources (OER) SHMS platform as a means of training. This is particularly evident when they possess an understanding of its advantages and exhibit a sense of ease in employing it to enhance their professional goals. Enhanced availability of technology and educational resources is crucial in establishing a favorable environment, particularly for educators to efficiently utilize the OER SHMS platform for resource discovery and utilization. To promote the use of Open Educational Resources (OER) in the field of education, it is imperative for administrators to establish conducive circumstances. This involves providing expert support for the development of new Open Educational Resources (OER) as well as the modification and contextualization of pre-existing OER. Furthermore, it is vital to advocate for the utilization of Open Educational Resources (OER) repositories, establish national OER repositories, implement quality assurance measures for OER, and effectively evaluate the endeavors of educators using OER into classroom teaching.

References

1. Abbas, S., et al.: Antecedents of trustworthiness of social commerce platforms: a case of rural communities using multi group SEM & MCDM methods. Electron. Commerce Res. Appl., 101322 (2023)
2. Al-Abdullatif, A.M., Alsubaie, M.A.: Using digital learning platforms for teaching Arabic literacy: a post-pandemic mobile learning scenario in Saudi Arabia. Sustainability **14**, 11868 (2022)
3. Al-Abrrow, H., Fayez, A.S., Abdullah, H., Khaw, K.W., Alnoor, A., Rexhepi, G.: Effect of open-mindedness and humble behavior on innovation: mediator role of learning. Int. J. Emerging Markets (2021)
4. Alasmari, T.M.: Mobile learning technology acceptance among Saudi higher education students. Wayne State University (2017)
5. Albahri, A.S., et al.: Based on the multi-assessment model: towards a new context of combining the artificial neural network and structural equation modelling: a review. Chaos Solitons Fractals **153**, 111445 (2021)
6. AL-Fatlawey, M.H., Brias, A.K., Atiyah, A.G.: The role of strategic behavior in achievement the organizational excellence" analytical research of the manager's views of Ur State Company at Thi-Qar Governorate". J. Adm. Econ. **10**(37) (2021)
7. Alghamdi, A.M., Alsuhaymi, D.S., Alghamdi, F.A., Farhan, A.M., Shehata, S.M., Sakoury, M.M.: University students' behavioral intention and gender differences toward the acceptance of shifting regular field training courses to e-training courses. Educ. Inf. Technol. (Dordr) **27**(1), 451–468 (2022)

8. Al-Hchaimi, A.A.J., Sulaiman, N.B., Mustafa, M.A.B., Mohtar, M.N.B., Hassan, S.L.B.M., Muhsen, Y.R.: Evaluation approach for efficient countermeasure techniques against denial-of-service attack on MPSoC-Based IoT using multi-criteria decision-making. IEEE Access **11**, 89–106 (2022)
9. Al-Mamary, Y.: Understanding the use of learning management systems by undergraduate university students using the UTAUT model: Credible evidence from Saudi Arabia. Int. J. Inf. Manage. Data Insights, 2 (2022)
10. Alnoor, A., et al.: How positive and negative electronic word of mouth (eWOM) affects customers' intention to use social commerce? A dual-stage multi group-SEM and ANN analysis. Int. J. Hum.–Comput. Interact., 1–30 (2022)
11. Alsalem, M.A., et al.: Rise of multiattribute decision-making in combating COVID-19: a systematic review of the state-of-the-art literature. Int. J. Intell. Syst. **37**(6), 3514–3624 (2022)
12. Al-Shamrani, A.: The ability of faculty members to use the Shams platform in Saudi universities. J. Educ. Psychol. Sci. **3**(28), 96–130 (2019)
13. Alyami, H.Y.: Integration of open educational resources in higher and general education institutions: from the perspectives of specialized and concerned bodies in E-Learning. World J Educ **10**(1), 30–41 (2020)
14. Antonietti, C., Cattaneo, A., Amenduni, F.: Can teachers' digital competence influence technology acceptance in vocational education? Comput. Hum. Behav., 132 (2022)
15. Arif, M., Ameen, K., Rafiq, M.: Factors affecting student use of Webbased services: application of UTAUT in the Pakistani context. Electr. Lib. **36**, 518–534 (2018)
16. Atiyah, A.G.: The effect of the dimensions of strategic change on organizational performance level. PalArch's J. Archaeol. Egypt/Egyptol. **17**(8), 1269–1282 (2020)
17. Atiyah, A. G. Strategic Network and Psychological Contract Breach: The Mediating Effect of Role Ambiguity (2023)
18. Atiyah, A.G., Zaidan, R.A.: Barriers to using social commerce. In: Artificial Neural Networks and Structural Equation Modeling: Marketing and Consumer Research Applications (pp. 115–130). Singapore: Springer Nature Singapore. https://doi.org/10.1007/978-981-19-6509-8_7
19. Cai, H., Dong, H., Li, X., Wong, L.H.: Does teachers' intention translate to actual usage? Investigating the predictors of K-12 teachers' usage of open educational resources in China. Sustainability **15**(2), 1027 (2023)
20. Chaveesuk, S., Khalid, B., Bsoul-Kopowska, M., Rostańska, E., Chaiyasoonthorn, W.: Comparative analysis of variables that influence behavioral intention to use MOOCs. PLoS ONE **17**(4), e0262037 (2022)
21. Chew, X., Khaw, K.W., Alnoor, A., Ferasso, M., Al Halbusi, H., Muhsen, Y.R.: Circular economy of medical waste: novel intelligent medical waste management framework based on extension linear Diophantine fuzzy FDOSM and neural network approach. Environ. Sci. Poll. Res., 1–27 (2023)
22. Daniali, S., et al.: Exploring UTAUT model in mobile 4.5G service: moderating social–economic effects of gender and awareness. Soc. Sci. 11 (2022)
23. De Hart, K., Chetty, Y., Archer, E.: Uptake of OER by staff in distance education in South Africa. Int. Rev. Res. Open Dist. Learn. **16**, 18–45 (2015)
24. Doan, Q.M, Dao, T.Q.: Factors influencing the intention to use open educational resources of students at universities in economics and business administration in Vietnam. J. Econ. Sustain. Dev. **11**(6) (2020)
25. Fadhil, S.S., Ismail, R., Alnoor, A.: The influence of soft skills on employability: a case study on technology industry sector in Malaysia. Interdiscip. J. Inf. Knowl. Manag. **16**, 255 (2021)
26. Gatea, A.A., Marina, V.: Higher education funding in Iraq in terms of the experience of particular developed countries. Int. J. Adv. Stud. **6**(1), 8–17 (2016)
27. Hair, J.F., Risher, J.J., Sarstedt, M., Ringle, C.M.: When to use and how to report the results of PLS-SEM. Eur. Bus. Rev. **31**(1), 2–24 (2019)

28. Hair, J.F., Hult, G.T.M., Ringle, C.M., Sarstedt, M.: A Primer on Partial Least Squares Structural Equation Modeling (PLS-SEM). Sage, Thousand Oaks, CA (2017)
29. Hamid, R.A., et al.: How smart is e-tourism? A systematic review of smart tourism recommendation system applying data management. Comput. Sci. Rev. **39**, 100337 (2021)
30. Henseler, J., Ringle, C.M., Sarstedt, M.: A new criterion for assessing discriminant validity in variance-based structural equation modeling. J. Acad. Mark. Sci. **43**, 115–135 (2015)
31. Khaw, K.W., et al.: Modelling and evaluating trust in mobile commerce: a hybrid three stage Fuzzy Delphi, structural equation modeling, and neural network approach. Int. J. Hum.-Comput. Interact. **38**(16), 1529–1545 (2022)
32. Kim, M.: Uses of creative commons licenses. Open Source Business Resource. Technology Innovation Management Review (2008)
33. Kumar, A., Buragohain, D., Deka, M.: Open educational resources (OER) issues and recommendations. In book: Bridging Educational Divides: OER and MOOCs (pp. 90–98) Publisher: National Law University Delhi (2019)
34. Kurelovic, K.: Acceptance of open educational resources driven by the culture of openness. In: 14th International Technology, Education and Development Conference (INTED) At: Valencia, Spain (2020)
35. Li, Y., Min, Z.: A study on the influencing factors of continued intention to use MOOCs: UTAUT model and CCC moderating effect. Front. Psychol., 12 (2021)
36. Manhal, M., Al-khalidi, A., Hamad, Z.: Strategic network: managerial myopia point of view. Manage. Sci. Lett. **13**(3), 211–218 (2023)
37. Menzli, L., Smirani, L., Boulahia, J., Hadjouni, M.: Investigation of open educational resources adoption in higher education using Rogers' diffusion of innovation theory. Heliyon **8**(7) (2022)
38. Mohan, M.M., Upadhyaya, P., Pillai, K.R.: Intention and barriers to use MOOCs: an investigation among the post graduate students in India. Educ. Inf. Technol. **25**, 5017–5031 (2020)
39. Nandwani, S., Khan, S.A.: Teachers' intention towards the usage of technology: an investigation using UTAUT model. J. Educ. Soc. Sci. **2**(2), 21 (2016)
40. National Center for e-Learning. SHMS– Saudi OER Network (2017). https://SHMS.sa/learn-more/
41. Padhi, N.: Acceptance and usability of OER in Indian higher education: an investigation using UTAUT model. Open Praxis **10**(1), 55–65 (2018)
42. Rahmaningtyas, W., Mulyono, K.B., Widhiastuti, R., Fidhyallah, N.F., Faslah, R.: Application of UTAUT (Unified Theory of Acceptance and Use of Technology) to understand the acceptance and use of the e-learning system. Int. J. Adv. Sci. Technol. **29**(4), 5051–5060 (2020)
43. Ramayah, T., Cheah, J., Chuah, F., Ting, H., Memon, M.A.: Partial least squares structural equation modeling (PLS-SEM) using smartPLS 3.0 (2018)
44. SHMS. About SHMS (2019). https://shms.sa/learn-more/
45. Smirani, L., Boulahia, J.: Using the unified theory of acceptance and use of technology to investigate the adoption of open educational resources by faculty members. Int. J. Inf. Technol. **14**, 3201–3211 (2022)
46. Tang, H., Lin, Y.J., Qian, Y.: Improving K-12 Teachers' acceptance of open educational resources by open educational practices: a mixed methods inquiry. Educ. Tech. Res. Dev. **69**, 3209–3232 (2021)
47. Tang, H., Lin, Y.J., Qian, Y.: Understanding K-12 teachers' intention to adopt open educational resources: a mixed methods inquiry. Br. J. Edu. Technol. **51**(6), 2558–2572 (2020)
48. Tarhini, A., Arachchilage, N.A.G., Abbasi, M.S.: A critical review of theories and models of technology adoption and acceptance in information system research. Int. J. Technol. Diffus. (IJTD) **6**(4), 58–77 (2015)

49. Teo, T., Sang, G., Mei, B., Hoi, C.K.W.: Investigating pre-service teachers' acceptance of Web 2.0 technologies in their future teaching: a Chinese perspective. Interact. Learn. Environ. **27**(4), 530–546 (2019)
50. Tillinghast, B.: Using a technology acceptance model to analyze faculty adoption and application of open educational resources. Int. J. Open Educ. Resour., **4**(1) (2021)
51. Tlili, A., Jemni, M., Khribi, M.K., Huang, R., Chang, T.W., Liu, D.: Current state of open educational resources in the Arab region: an investigation in 22 countries. Smart Learn Environ. **7**(1), 1–15 (2020)
52. Tseng, T., Lin, S., Wang, Y., Liu, H.: Investigating teachers' adoption of MOOCs: the perspective of UTAUT2. Interact. Learn. Environ. **30**(4), 635–650 (2022)
53. Valtonen, T., Kukkonen, J., Kontkanen, S., Sormunen, K., Dillon, P., Sointu, E.: The impact of authentic learning experiences with ICT on pre-service teachers' intentions to use ICT for teaching and learning. Comput. Educ. **81**, 49–58 (2015)
54. Venkatesh, V., Morris, M.G., Davis, G.B., Davis, F.D.: User acceptance of information technology: toward a unified view. MIS Q., 425–478 (2003)
55. Weller, M., De Los Arcos, B., Farrow, R., Pitt, B., McAndrew, P.: The impact of OER on teaching and learning practice. Open Praxis **7**(4), 351–361 (2015)
56. Yeboah, D., Nyagorme, P.: Students' acceptance of WhatsApp as teaching and learning tool in distance higher education in sub-Saharan Africa. Cogent Educ. **9**, 2077045 (2022)
57. Zaidan, A.S., Arianpoor, A.: Artificial neural network and structural equation modeling techniques. In: Alnoor, A., Wah, K.K., Hassan, A. (eds.) Artificial Neural Networks and Structural Equation Modeling: Marketing and Consumer Research Applications, pp. 3–22. Springer Nature Singapore, Singapore (2022). https://doi.org/10.1007/978-981-19-6509-8_1

How Are Brand Activity and Purchase Behavior Affected by Digital Marketing in the Metaverse Universe?

Nadia Atiyah Atshan[1], Hasan Oudah Abdullah[2], Hadi AL-Abrrow[3], and Sammar Abbas[4](✉)

[1] Management Technical College, Southern Technical University, Basrah, Iraq
[2] Department of Business Administration, Basrah University College for Science and Technology, Basrah, Iraq
[3] Department of Business Administration, College of Administration and Economics, University of Basrah, Basrah, Iraq
[4] Institute of Business Studies, Kohat University of Science and Technology, Kohat, Pakistan
sabbas@kust.edu.pk

Abstract. With the widespread adoption of Metaverse for digital marketing purposes, a debate has flourished about the opportunities it offers to both customers and marketing companies. To address the issues relevant to digital marketing and the metaverse, a novel paradigm is proposed in this paper that investigates how the metaverse (novelty, interactivity, vividness) might impact digital marketing. The paper also investigates the mediating role of digital marketing in the relationship between brand activity and purchase behavior. The sample of this research included 350 clients selected from the hotel, restaurant, and service sectors who deal with services that are marketed according to metaverse tools. The study also provides a checklist for researchers to investigate the potential benefits of the relationship between the metaverse and digital marketing.

Keywords: Digital Marketing · Brand Activity · Purchase Behavior · Metaverse · Interactivity

1 Introduction

Digital marketing is the process of creating channels for potential customers using digital technologies to achieve business goals through the satisfaction of customers' needs. The term "digital marketing," has gained popularity over the past few years, and many concepts such as marketing" and "e-marketing" are occasionally used interchangeably. However, each one of these terms can carry a different meaning. For example, the Internet is just one of many possible channels to reach customers; it is not the only one. Additionally, there are home appliances, audio and video equipment that can be used to access the widely spread customers (Ning et al., 2023; Atiyah and Zaidan, 2022). The importance of digital markets to the modern economy and their effects on, among other things, privacy, political pluralism, and cultural diversity have placed them at the

M. Al-Emran et al. (Eds.): IMDC-IST 2024, LNNS 876, pp. 112–128, 2023.
https://doi.org/10.1007/978-3-031-51300-8_8

forefront of public policy discussions. Any argument for policy must include a thorough understanding of how these markets operate (Dubey et al., 2022). While there are currently 8 billion people on the planet, more than 5 billion of them are using the internet. This represents an increase of 62% from the year 2010 in the use of digital technologies. The top technology companies in the world, including Amazon, Google, Apple, and many others, benefited from this exceptional growth by starting businesses connected to digital marketing in one or another way at the start of online commerce (Ali & Khan, 2023). A recent development in digital marketing is known as "metaverse marketing." The metaverse is a virtual reality environment geared toward social interaction and accessed through virtual reality (VR) and augmented reality (AR) headsets (Dwivedi et al., 2022).

Many businesses have started to adopt the transformation that the Metaverse brought about. Businesses have made a lot of announcements about their plans for the Metaverse, and this practice is becoming a usual one. Metaverse has developed into a setting where businesses can carry out commercial operations virtually. A growing number of companies are using Metaverse for online meetings with remote employees. The employees may participate in the sessions using digital avatars. In addition, segmenting the market allows for a variety of business interactions with clients (Zhang et al., 2022). Beyond embracing and utilizing contemporary technology, marketing success in the current digital era is largely dependent on the efficient utilization of digital technologies. Maintaining current marketing trends with emerging trends like metadata storage is also important. Metaverse is a permanent, 3D virtual environment where users can hang out while being targeted by brand-related content and sales activation strategies in the marketing context (Tlili et al., 2022; Hadi et al., 2018). Businesses using immersive technologies are predicted to significantly alter consumer behavior in the Metaverse. Businesses' entry into the Metaverse will be accelerated by adhering to a specific Metaverse planning. Consumer behavior will vary depending on the approach businesses take to adopt to metaverse. The as a result, the Metaverse determines how virtual world elements and consumer behavior are exchanged (Mystakidis, 2022).

The goal of Metaverse's digital marketing strategy is to benefit customers who prioritize innovation and exclusive access to products using technologies that combine virtual reality, augmented reality, mixed reality, and artificial intelligence. Brands can fully reach their target audience persuasively and creatively through metaverse marketing. As a result, brands are increasingly focusing on digital marketing in the Metaverse (Kye et at., 2021). Along with consumer behavior changes brought on by Metaverse, brands entering Metaverse is also appreciated. This study seeks to add to the literature by thoroughly understanding the concept of the Metaverse, which is quickly growing in popularity and usage in the context of digital marketing. It aims to serve as a starting point for brands as they begin their marketing studies.

2 Literature Review

2.1 Metaverse

Metaverse is a perpetual and persistent multiuser environment that combines physical reality with virtuality. As a backdrop, the post-reality era serves as the backdrop. Multisensory interactions with digital objects, virtual environments, and people are made possible by the fusion of technologies, including virtual reality (VR) and augmented reality (AR) (Dwivedi et al., 2022; Alnoor et al., 2022a). A persistent, multi-user, socially networked platform web is known as the Metaverse. Digital artifacts can be dynamically interacted with the seamless embodied user communication (Mystakidis, 2022). Its original form was a network of virtual worlds connected by teleportation. Cooperative AR environments, social, immersive VR platforms that support open game worlds, and massively multiplayer online games are all present in the modern Metaverse (Anderson & Rainie, 2022). It allows for seamless transmission of an embodied tangible moment while allowing for dynamic user interaction without visible digital objects. When it first launched, interactions were the only teleportation route that avatars could use to travel between virtual worlds (Wang et al., 2022a).

Known as a next-generation Internet paradigm, the metaverse is where users can experience virtual lives and live as digital natives following the web and mobile Internet revolutions (Setiawan & Anthony, 2022). An imagined metaverse is a virtual shared space that combines the three realms of the physical, the human, and the digital. It is fully immersive, hyper-spatiotemporal, and self-sustaining (Dubey et al., 2022). Multisensory interactions with digital characters, objects, and environments are possible in the metaverse thanks to technologies. The XR system can be made more accurate by using separate displays for each eye that are slightly different and mimic sight in actual environments. The stereoscopic display conveys a sense of depth and is able to give the XR system high representational fidelity (Tlili et al., 2022; Alnoor et al., 2022b). When compared to 2D systems, XR systems offer better auditory quality. By creating soundscapes with binaural, 3D, or spatial audio, AR and VR systems greatly enhance immersion (Dubey et al., 2022). The spatial distribution of sound makes it an effective navigation tool and attention magnet by allowing users to recognize the directions of sound cues (Kye et al., 2021).

Since different ICTs coexist harmoniously in the convergent universe, Metaverse is not a stand-alone technology. As an alternative, Metaverse can combine several ICTs currently in use, allowing users to use them all at once. Now users can network and socialize in an interoperable setting. The platform-independent nature of accounts and profiles enables users to connect and navigate across virtual worlds (such as Horizon World, Sandbox, and Roblox). This generates unheard-of marketing opportunities (Lee, 2021). Users' vivid imaginations can be strengthened because Metaverse can support the development of a strong sense of presence and embodiment. So, it stands to reason that Metaverse will be more effective at inspiring fantasy and imagination (Zhao et al., 2022). Due to the Metaverse's potential for business, numerous multinational corporations have already started investing in it. For instance, J.P. Morgan started offering services to customers in the Metaverse in February 2022 after opening its first bank branch there (Shevlin, 2022). High-end companies like Ralph Lauren and Gucci also created virtual stores to sell digital clothing on Roblox. In parallel, hotel chains like Citizen M and EV Hotel Corporation have begun to build virtual hotels in Sandbox so that they can interact with other hotel guests (Wong, 2022).

2.2 Opportunities of Metaverse

The metaverse's extensive features and resources allow businesses to market goods that would be impractical in the real world. Because there are no natural laws in the virtual universe, sellers are free to use their creative minds to develop goods that are different from those found in the real world (Gursoy et al., 2022). Due to the virtual nature of the metaverse, brands will be able to engage with a wider variety of customers with greater levels of immersion. While it is impossible in other forms of media, advertisements in the metaverse can be very interactive (Bala & Verma, 2018). Ads can be made with elements and information that are ethereal and transcendent, providing users with an experience that is unreal. Due to the high level of involvement and communication, clients will be able to use the products, enabling brands to create more dependable buying habits. Increased adoption, accessibility, and cost savings in the metaverse will change consumer behavior and experiences significantly (Zhao et al., 2022). Similarly, the adoption of creativity theory determines that given the growing popularity of the metaverse in terms of communication channels, personal and professional interactions, the platform's interface and integration, and continual attempts to improve the real-world experience, consumer segments will join it (Wong, 2022).

Two subcategories of the metaverse: "metaverse as a tool" and "metaverse as a target." The phrase "using the metaverse as a tool" refers to the application of the metaverse to problems and issues in the physical universe. The phrase "metaverse as a target" refers to the metaverse's capacity to carry out tasks like metaverse expansion and revenue generation. Applications that use the metaverse as a target, as opposed to a tool, are autonomous and heavily reliant on the virtual world (Zhao et al., 2022).

Metaverse is an intriguing new market for businesses. Different business models are used by organizations in the Metaverse to generate revenue and enhance promotion activities. In comparison to virtual products, real ones can be produced with fewer steps and resources. For customers, the input interface in the metaverse can be made simpler (e.g., by using a kiosk) (Duan et al., 2021). In the metaverse, social issues, moral conundrums, and policy-related issues can all be simulated without bias or social discrimination. The metaverse facilitates the social and marketing phenomena. Because client experience analysis is more accurate than survey analysis based on user opinions, the metaverse is more suitable for business usage. Although it is a common misconception that people drink more water than they do soda, accurate research can be aided by user logs that are stored in the metaverse. Additionally, quantitative metaverse data is used to predict user behavior (Narin, 2021). The metaverse provides an opportunity to encounter dangerous environments. (For instance, educating children about outdoor safety and providing transportation for seniors using public transportation.). In terms of interactive learning, it is more effective than audiovisual education. (Such as ray leaks, conflict areas, and crisis response). The metaverse has made it possible for more people to access educational opportunities (Lim et al., 2022).

3 Hypothesis Development

3.1 Relationship Between Digital Marketing and Metaverse

Digital marketing campaigns use online platforms like websites, social media sites, email, search engines, pop-up advertisements, and frequently targeted intrusive advertisements delivered by cookies on electronic devices like desktop computers, mobile devices, and intelligent devices to consume media. Digital marketing is the term used to describe targeted, quantifiable, and engaging promotion of goods and services to acquire, convert, and retain customers. (Dwivedi et al., 2020; AL-Fatlawey et al., 2021). In recent years, the world has changed, moving from print media to email marketing to social media marketing. According to estimates, the global digital marketing market was worth $350 billion in 2020 and is expected to be $786.2 billion in 2026 (Lee et al., 2021). Be it a small or large company, marketing is a crucial pillar for success. The success or failure of any company can be directly attributed to its marketing strategies, which are directly aligned with its vision and mission. The world will eventually transition into the Metaverse as a whole, it is not incorrect to predict, and social media platforms will soon become obsolete (Morris, 2009). Marketers have fully embraced the data-driven strategy approach. Every effective marketing campaign uses data-driven and user-centric strategies to achieve its goals. Businesses can gather more precise and trustworthy data from their users in the presence of technologies like augmented reality, virtual reality, artificial intelligence, etc. This will enable them to satisfy all their customers' needs (Nalbant & Aydin, 2023). The most advanced form of digital marketing that has ever been used is predicted to be metaverse marketing. Marketers are taking the great initiative to develop effective marketing strategies and provide their users with a fantastic real-like virtual experience. They will stop at nothing to establish their brand as distinctive and well-known in the Metaverse (Ali & Khan, 2023).

Success in marketing in the modern digital era involves more than just embracing and utilizing new technology. An integration of existing technologies with emerging trends like metadata storage is important (Hollensen et al., 2022). To give consumers who value innovation exclusive access, Metaverse advises using technologies that combine virtual reality, augmented reality, mixed reality, and artificial intelligence as part of their digital marketing strategy (Efendioglu, 2023). The most effective way to engage with their target audience is by using metaverse marketing, which allows brands to unleash their creativity. Brands are consequently increasingly relying on Metaverse for digital marketing. In addition to realizing the importance of digitalization, businesses are developing their strategies for the virtual market (Chew et al., 2023; Abbas et al., 2023). To rule the virtual markets, the established and conventional methods must be modified. Resolving issues with consumer trust is the first challenge that arises in this situation (Rathore, 2018). The visual component is another crucial component of selling in the metaverse. To maintain interest in the virtual world of the metaverse, the user must be totally immersed. The aesthetic appeal will increase sales in the metaverse of both digital and physical products. The companies are using the "vault" sandbox to create their own metaverse, taking inspiration from the prestigious clothing brand Gucci (Dwivedi et al., 2023). Businesses that rely on customers' physical experiences can profit from SEOs in the metaverse. As a trend, the metaverse will be used by an increasing number of businesses involved in the fashion industry Gucci is preparing to be included in the metaverse and has already established itself in the AR world (Hennig-Thurau & Ognibeni, 2022). In addition, Manufacturing facilities, gigafactories, warehouses, and supply chains will become unnecessary in the markets where digital merchandising is essential. The economy will be directed to avatar D2A, and a digital twin of the real-world item—such as clothes, accessories, and handbags for the avatar—will be instantly made available in the metaverse. Additionally, since digital products will be distributed using blockchain technology, there won't be any chance of fakes or copies (Garda, 2023).

Customers can enjoy a consistent, one-of-a-kind experience because of augmented reality as it merges the real world and the digital one. When using an augmented reality feature, individuals are likely to be exposed to fresh stimuli due to the breadth and depth of manipulation between the real and virtual worlds. Thus, the novelty here refers not to the "newness" of augmented reality but also to the novel, special, individual, and creative content (stimuli) that the AR display is continuously exposed to (Tan et al., 2023). Virtual furnishings can be placed in real spaces by AR users. Thanks to the unique way this content is presented, which results in new, personalized content, users can visualize how a piece of furniture will look in their homes. Due to this, augmented reality (AR) allows users to alter the content to match their preferences and interests (Lamba & Malik, 2022). Furthermore, Technological elements that involve speed, like how quickly users can edit data, could influence how consumers view products. The interaction includes one's individual opinions of others i.e., the viewpoint of the user regarding interaction (Cheah & Shimul, 2023). Interactivity, in particular, aids in the provision of a thorough description that makes clear the function of interactivity in augmented reality due to technology and user perception. Thus, the technological system's capacity to encourage user involvement and participation in digital marketing results in interactivity (Aliev & Kadirov, 2022). The quality of the information displayed can be improved while the number of sensory dimensions is increased. A vivid environment will influence the way psychological characteristics are elaborated and enhance the recall of previously retained information. This may completely or adversely affect consumers' preferences for products, depending on the value of the knowledge that is remembered. Customers can visualize upcoming interactions with a product in their minds thanks to vividness, just like with interactivity (Rani, K., & Singh, 2023).

The first hypothesis of the study in this regard is provided below:

H1a: There is a significant influence of novelty in metaverse over digital marketing.

H1b: There is a significant influence of interactivity in metaverse over digital marketing.

H1c: There is a significant influence of vividness in metaverse over digital marketing.

3.2 Digital Marketing as a Mediator Variable

The use of the Internet and other interactive media to start and connect conversations between companies and particular target audiences is known as digital marketing. When creating strategies to access customers and guide them toward a combination of electronic communication and traditional communication, keeping up with developments in digital technology is essential (Indumathi, 2018; Albahri et al., 2022). Because consumers frequently make comparisons between products and the creator's brand. When products are viewed as having value and having an impact on a company's brand, the power of a brand can support the value of those products. It is obvious that the use of digital marketing affects brand activity, which is evaluated by consumers and is based on their preferences (Alamsyah et al., 2021).

Digital marketing usage has changed how customers interact with brands. The economics of marketing are changing because of the obsolescence of many traditional strategies and structures used in the function. Marketers can no longer succeed by operating in the traditional manner (Melović et al., 2020), where users of digital marketing today interact with their favorite brands as well as one another. This has allowed businesses to communicate with and learn more about their customers directly (Purwanto, 2022). Using digital marketing, businesses can connect with their customers and persuade them to purchase a particular product, learn about a particular brand, or leave reviews about that brand on various social media platforms. Effective social media marketing requires listening, understanding, and some form of customer participation to add value and strengthen business relationships, rather than just grabbing the customer's attention to deliver a message or elicit a response (Satria & Hasmawaty, 2021; Alnoor et al., 2022a).

Through product browsing and reviews, digital marketing offers consumers a variety of settings to direct and control the purchasing processes. Without adequate marketing and product content to persuade and convince customers for a potential sale and to ensure repeat purchases, sales conversion is a challenging process (Al-Azzam & Al-Mizeed, 2021; Atiyah, 2022). Digital marketing, one of the most innovative marketing techniques, is frequently used to interact with customers and advertise products and services regardless of place, time, or cost. The appealing contents of digital marketing and the customization for the customers had a significant and positive influence on purchase intention (Sigar et al., 2021). An effective digital marketing is based on superior performance over a specific period, good digital marketing will easily create good customer relationships (Dar & Tariq, 2021). So, Purchase decision refers to a customer's intention to use a brand's product or service, as well as a customer's desire to do so. Customers' purchase intentions are influenced by the level of satisfaction they expect and receive (Khwaja et al., 2020). Consumers are more likely to act quickly and make more purchases when they receive digital marketing messages about products or services (Hien & Nhu, 2022). The hypothesizes of the study in this regard are provided below:

H 2: There is a positive effect of digital marketing on brand activity.

H 3: There is a positive effect of digital marketing on purchase behavior.

H 4: Digital marketing mediates the relationship between the metaverse and brand activity.

H 5: Digital marketing mediates the relationship between the metaverse and purchase behavior.

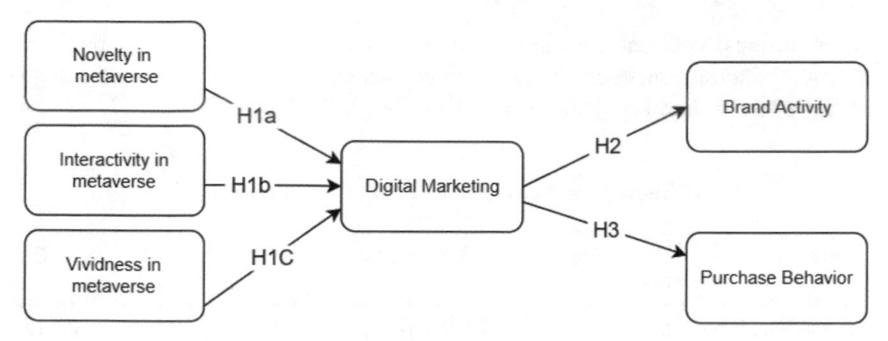

Fig. 1. Research Framework (authors).

4 Methodology

4.1 Sample Size and Measurements

The sample of this study included 350 clients who were engaged with digital marketing activities applying metaverse tools. Overall, 400 customers were reached, and 350 final responses were considered for the study resulting in a response rate of 87.5%. Participants are selected from the hotel, restaurant, and service sectors through purposive sampling. The reason for this is that in those sectors there are services that are provided and marketed in a modern way. Participants were selected based on the following criteria: a) who experienced seeing marketing services via metaverse, and b) who experienced those services. Ready-made measures were used for all variables except for the brand activity variable, whose measure was developed through the current study using four items. The measures can be clarified as follows:

Metaverse: A scale developed by Yim et al. (2017) was used that included three dimensions (Vividness, Interactivity, Novelty) and thirteen items. Six items for the vividness dimension, three items for the interactivity dimension, and four items for the novelty dimension.

Digital marketing: for this a scale of Yadav & Rahman (2017) was used that includes five dimensions (Interactivity, Informativeness, Personalization, Trendiness, Word of Mouth) and fifteen items divided equally.

Purchase Behavior: Based on Omar and Atteya, (2020), a four-item scale was used to measure this variable. Scales are fully shown in the appendices.

4.2 Statistical Approach

To test the proposed hypotheses and analyze data we relied on AMOS version 25 and SPSS version 25. The validity and reliability of the measures were confirmed, and then descriptive statistics and correlation coefficients were presented.

5 Results

Before hypothesis testing, a confirmatory factor analysis (CFA) was performed to assess the factor structure of the self-report items that measured the four central constructs in this study: metaverse, digital marketing, brand activity, and purchase behavior (Hair et al.,

2014). Additionally, Cronbach's alpha reliability was examined to gauge the internal consistency of these constructs. Notably, all constructs exhibited values exceeding 0.7, indicating strong reliability (Hair et al., 2014) (See Table 1).

Table 1. Reliability and Validity Measurement.

Construct	Number of items	Range of Loadings	α	C.R	AVE
Metaverse Vividness	6	0.599–0.876	0.876	0.890	0.546
Metaverse Interactivity	3	0.643–0.887	0.798	0.889	0.554
Metaverse Novelty	4	0.578–0.905	0.763	0.769	0.509
Digital marketing: Interactivity	3	0.705–0.894	0.808	0.907	0.654
Digital marketing: Informativeness	3	0.765–0.879	0.775	0.902	0.654
Digital marketing: Personalization	3	0.732–0.799	0.798	0.879	0.621
Digital marketing: Trendiness	3	0.689–0.897	0.805	0.895	0.609
Digital marketing: Word of Mouth	3	0.669–0.923	0.776	0.809	0.621
Brand Activity	4	0.587–0.976	0.765	0.799	0.567
Purchase Behavior	4	0.632–0.889	0.846	0.867	0.582

To establish convergent validity, the average variance extracted (AVE) and composite reliability (CR) were evaluated (Fornell and Larcker, 1981). Both AVE and CR values surpassed the threshold of 0.5, underscoring robust convergent validity (Table 2).

Discriminant validity was then assessed by computing the square root of the AVE for each construct (Refer to Table 2) and comparing these values with the inter-construct correlations (Fornell and Larcker, 1981). This analysis unequivocally confirmed the discriminant validity of the constructs.

Lastly, a common method bias (CMB) assessment was performed using the Harmon single-factor technique. Findings indicated that this index yielded a value of 0.33, significantly lower than the threshold value of 0.5 (Zhou and Long, 2004). This outcome affirms the absence of substantial common method bias in the study's results.

Table 2 presents the correlation matrix (including demographic variables). It can be observed that the independent variables, mediating variables, and dependent variables are significantly correlated with each other. In addition, demographic variables like education and gender of the current supervisor are significantly correlated with brand activity and purchase behavior.

Table 2. Means, Standard Deviations, and correlations among the Variables.

	Gender	Education	Age	VIV	INT	NOV	DM	BA	PB
Gender	1								
Education	.054	1							
Age	.032	.023	1						
VIV	.002	.010	.012	1					
INT	.054	.011	.025	.345**	1				
NOV	.041	.035	.019	.440**	.337**	1			
DM	.032	.018	.009	.139*	.210*	.198*	1		
BA	.167*	.159*	.008	.654**	.309**	.289**	.564**	1	
PB	.232*	.187*	.06	.546**	.229**	.456**	.632**	.455**	1

Notes: * $p < 0.05$; ** $p < 0.01$; VIV = Metaverse Vividness; INT = Metaverse Interactivity; NOV = Metaverse Novelty; DM = Digital marketing; BA = Brand Activity; PB = Purchase Behavior.

Table 3 summarizes the various relationships between the variables under study. Table 3 explains the relationships between variables and the mediation relationship between the variables.

Table 3. Hypothesis testing.

Hypothesis	Path	β	S.E	T-value	P-value	Result
H1a	VIV → DM	.334	.023	11.189	0.000	Accepted
H1b	INT → DM	.432	.019	19.404	0.000	Accepted
H1c	NOV → DM	.336	.032	7.167	0.000	Accepted
H2	DM → BA	.543	.033	13.122	0.000	Accepted
H3	DM → PB	.550	.026	17.821	0.000	Accepted
H4a	VIV → DM → BA	.181	.018	6.723	0.000	Accepted
H4b	INT → DM → BA	.235	.015	12.334	0.000	Accepted
H4c	NOV → DM → BA	.182	.014	9.667	0.000	Accepted
H5a	VIV → DM → PB	.184	.013	10.821	0.000	Accepted
H5b	INT → DM → PB	.238	.011	18.303	0.000	Accepted
H5c	NOV → DM → PB	.185	.027	3.519	0.002	Accepted

Through the results in the table above, all hypotheses in the study are accepted. The results indicate the acceptance of the first hypothesis, which indicates that there is a positive effect of vividness, interactivity, and novelty on digital marketing. The second and third hypotheses were also accepted, which indicate that there is a positive impact relationship of digital marketing on brand activity and purchase behavior. Finally, the results also indicate the acceptance of the hypotheses of indirect influence, i.e., that vividness, interactivity, and novelty affect brand activity and purchase behavior through digital marketing.

6 Implications and Future Studies

6.1 Theoretical Implications

The current study answers questions about what metaverse are and how they affect digital marketing in a competitive environment, giving answers to the researchers who are interested in metaverses and its potential to link to digital marketing. By identifying and categorizing various determinants of metaverse with digital marketing, this research study has added new knowledge to the field of study. Previous researchers used a variety of variables to pinpoint these two activities in various contexts and circumstances. This study's definition of the metaverse as a fresh approach to enhancing marketing represents a significant advancement in the field of study. Additionally, for the first time, these aspects of the metaverse were used to research how they affected digital marketing. We studied the relationship between metaverse dimensions and digital marketing. The findings indicated that novelty, interactivity, and vividness have an impact on digital marketing. There is a link between digital marketing, brand activity, and purchase behavior. This is a significant contribution to determining the antecedents of the metaverse.

6.2 Practical Implications

Digital sharing on online platforms is becoming essential to almost every aspect of our lives in the rapidly developing digital world. The metaverse's introduction has sped up the adoption of new technologies and technology-based services. The metaverse has increased the number of first-time online buyers as well as the number of personal and nearly professional jobs that people must perform. Another essential component is consumers' readiness and willingness to use the Metaverse as their main platform for interacting with marketing destinations and organizations, looking for and buying marketing goods, and creating value with other users. Online marketing destinations and businesses must be prepared to welcome and assist virtual clients on Metaverse platforms, as well as to deliver a satisfying user experience and enough value. Furthermore, electronic advertising and the metaverse environments in the physical and virtual worlds can interact and affect one another without posing a risk to privacy and security thanks to extensive edge intelligence services. Lastly, this study presents a framework for controlling digital marketing activity and directs marketers to effectively use social media for brand activity and consumer behavior. To participate in the metaverse, marketers can now access consumer information on what they value most. Thus, by gaining a thorough understanding of the brand and consumer behavior, this research equips marketers with strategies for effectively participating in the metaverse for digital marketing.

6.3 Future Studies

The future belongs to digital technologies when these technologies will overcome the traditional ways of doing business. In recent times we have witnessed a greater penetration of digital technologies into every sphere of life. The emergence of the Metaverse has further signified the role of digital technologies and offered novel opportunities for businesses to transform their operations. This also calls for research efforts to further investigate the role of digital technologies in the context of Metaverse. The current study investigates the role of digital marketing in influencing brand activity and purchase behavior in the presence of Metaverse in the context of the hotel industry. Future studies can take into account other businesses like manufacturing ones to look at the role of digital marketing and metaverse. This study has considered three aspects of the metaverse (novelty, interactivity, and vividness). Future studies may take into account other components of the metaverse such as corporeity, persistence, immersion, advanced computing, socialization and decentralization.

7 Conclusion

The way we interact, experience, and interact with the digital world could be fundamentally changed by the metaverse. It integrates a number of cutting-edge technologies into a virtual reality setting in order to give users an immersive and interactive experience, including 5G communication, brain-computer interfaces, virtual reality, blockchain, cloud computing, digital twins, and artificial intelligence. Thanks to the exponential growth of the metaverse, brands, and marketers now have new opportunities to develop cutting-edge strategies for reaching their target markets. Planning for the long term and developing strategies that can incorporate the Metaverse in the traditional society, where all norms and values are upheld for a sustainable environment, are urgently needed. At a breakneck pace, technology is developing. Customers will also engage in an immersive buying or selling experience when the imaginative and inventive virtual marketing strategy known as metaverse marketing is used. And they are currently utilized in different industries, businesses, and everyday activities.

Understanding the metaverse and its full potential is one of the many technological advancements that digital marketers must stay on top of. Marketers need to understand that the metaverse offers a chance to create marketing experiences that relate to real-world activities or imitate what businesses already do in the real world. The metaverse offers businesses ample opportunity for brand development and marketing. Metaverse marketing experimentation, development, and success are unlikely to be significantly hampered by the limitations of current technology and the low level of public acceptance.

The findings demonstrate how the idea of digital marketing in the metaverse can be applied as a marketing tool to comprehend a range of digital marketing factors and circumstances. In other words, the research sample relied on it to elevate digital marketing to the top of its industry. Consequently, the market is expanding quickly. Due to the use of digital marketing in the metaverse, it has an impact on consumer behavior and brand recognition, driving consumers to seek out the company's products.

Appendices: Study Scale

Metaverse: (Yim et al., 2017)
Metaverse Vividness

1. The visual display & Audio through the Metaverse technology was clear.
2. The visual display & Audio through the Metaverse technology were sharp.
3. The visual display and audio through the Metaverse technology were detailed.
4. The visual display and audio through the Metaverse technology were vague.
5. The visual display and audio through the Metaverse technology were vivid.
6. The visual display and audio through the Metaverse technology were well-defined.

Metaverse Interactivity

7. You were in control of communicating with brands through Metaverse technology.
8. You have some control of the Brand's Metaverse technology.
9. The brands in Metaverse can respond to your specific needs quickly and efficiently.

Metaverse Novelty

10. Using the Metaverse Technology feature offers something new each time.
11. Using the Metaverse Technology feature offers unique information.
12. Using the Metaverse Technology features something different each time for me.
13. Using the Metaverse Technology feature offer specific content.

Digital marketing: (Yadav & Rahman, 2017)
Interactivity

1. social media for e-commerce enables me to share and update the current content.
2. This online retailer communicates with its customers frequently.
3. social media for e-commerce makes it easier to have two-way conversations with loved ones.

Informativeness

1. social media for e-commerce provides accurate product information.
2. social media for e-commerce provides helpful information
3. E-commerce's social media platforms offer thorough information.

Personalization

1. social media for e-commerce recommends products based on my needs
2. I believe that using social media for e-commerce satisfies my needs.
3. social media for e-commerce makes it easier to find personalized information

Trendiness

4. The newest trend is content that is accessible on social media for e-commerce.
5. Using social media for e-commerce is hip.
6. social media for e-commerce offers everything that is in style.

Word of Mouth

7. I would advise my friends to check out social media for e-commerce.
8. I would recommend e-commerce social media to my friends and acquaintances.
9. On e-commerce social media, I want to tell my friends and acquaintances about my shopping experiences.

Brand Activity (Developed for this study)

1. I am very present on the social networking sites of my brand page that I deal with.
2. I share content that I like on my brand's social media sites, which I deal with on an ongoing basis.
3. It took me a short time to create content that supported my brand.
4. I participate in events and events that my brand creates on social media.

Purchase Behavior (Omar & Atteya, 2020)

1. I would prefer making an online purchase rather than going to the outlet.
2. I would prefer to buy from a store where I may get better offers, like free home delivery.
3. Shopping online is too time-consuming.
4. I would compare the performance of the product(s) with that of the promise made by the website.

References

Abbas, S., et al.: Antecedents of trustworthiness of social commerce platforms: a case of rural communities using multi group SEM & MCDM methods. Electron. Commerce Res. Appl., 101322 (2023)

Alamsyah, D.P., Ratnapuri, C.I., Aryanto, R., Othman, N.A.: Digital marketing: implementation of digital advertising preference to support brand awareness. Acad. Strateg. Manag. J. **20**(2), 1–10 (2021)

Al-Azzam, A.F., Al-Mizeed, K.: The effect of digital marketing on purchasing decisions: a case study in Jordan. J. Asian Financ. Econ. Bus. **8**(5), 455–463 (2021)

Albahri, O.S., et al.: Novel dynamic fuzzy decision-making framework for COVID-19 vaccine dose recipients. J. Adv. Res. **37**, 147–168 (2022)

AL-Fatlawey, M.H., Brias, A.K., Atiyah, A.G.: The role of strategic behavior in achievement the organizational excellence. Analytical research of the manager's views of Ur State company at Thi-Qar governorate. J. Adm. Econ. **10**(37) (2021)

Ali, S.A., Khan, R.: Metaverse marketing vs digital marketing. Int. J. Innov. Sci. Res. Technol. **8**(1), 385–388 (2023)

Aliev, A., Kadirov, D.: Digital marketing and smart technology marketing systems as the future of metaverse. In: Koucheryavy, Y., Aziz, A. (eds.) International Conference on Next Generation Wired/Wireless Networking, pp. 397–410. Springer, Cham (2022). https://doi.org/10.1007/978-3-031-30258-9_35

Alnoor, A., et al.: Uncovering the antecedents of trust in social commerce: an application of the non-linear artificial neural network approach. Competitiveness Rev. Int. Bus. J. **32**(3), 492–523 (2022)

Alnoor, A., Khaw, K.W., Al-Abrrow, H., Alharbi, R.K.: The hybrid strategy on the basis of Miles and Snow and Porter's strategies: an overview of the current state-of-the-art of research. Int. J. Eng. Bus. Manag. **14**, 18479790221080216 (2022)

Alnoor, A., et al.: How positive and negative electronic word of mouth (eWOM) affects customers' intention to use social commerce? A dual-stage multi group-SEM and ANN analysis. Int. J. Hum. Comput. Interact., 1–30 (2022)

Anderson, J., Rainie, L.: The metaverse in 2040. Pew Research Centre, 30 (2022)

Atiyah, A.G.: Effect of temporal and spatial myopia on managerial performance. Journal La Bisecoman **3**(4), 140–150 (2022)

Atiyah, A.G., Zaidan, R.A.: Barriers to using social commerce. In: Alnoor, A., Wah, K.K., Hassan, A. (eds.) Artificial Neural Networks and Structural Equation Modeling: Marketing and Consumer Research Applications, pp. 115–130. Springer, Singapore (2022). https://doi.org/10.1007/978-981-19-6509-8_7

Bala, M., Verma, D.: A critical review of digital marketing. Int. J. Manag., IT Eng. **8**(10), 321–339 (2018)

Cheah, I., Shimul, A.S.: Marketing in the metaverse: moving forward–what's next? J. Glob. Scholars Market. Sci. **33**(1), 1–10 (2023)

Chew, X., Khaw, K.W., Alnoor, A., Ferasso, M., Al Halbusi, H., Muhsen, Y.R.: Circular economy of medical waste: novel intelligent medical waste management framework based on extension linear Diophantine fuzzy FDOSM and neural network approach. Environ. Sci. Pollut. Res., 1–27 (2023)

Dar, T.M., Tariq, N.: Footprints of digital marketing on customers' purchase decision. Electron. Res. J. Soc. Sci. Hum. **3**, 20–30 (2021)

Duan, H., Li, J., Fan, S., Lin, Z., Wu, X., Cai, W.: Metaverse for social good: a university campus prototype. In: Proceedings of the 29th ACM International Conference on Multimedia, pp. 153–161, October (2021)

Dubey, V., Mokashi, A., Pradhan, R., Gupta, P., Walimbe, R.: Metaverse and Banking Industry–2023 The Year of Metaverse Adoption (2022)

Dwivedi, Y.K., et al.: Metaverse beyond the hype: multidisciplinary perspectives on emerging challenges, opportunities, and agenda for research, practice, and policy. Int. J. Inf. Manage. **66**, 102542 (2022)

Dwivedi, Y.K., et al.: Metaverse marketing: how the metaverse will shape the future of consumer research and practice. Psychol. Mark. **40**(4), 750–776 (2023)

Dwivedi, Y.K., Rana, N.P., Slade, E.L., Singh, N., Kizgin, H.: Editorial introduction: advances in theory and practice of digital marketing. J. Retail. Consum. Serv. **53**, 101909 (2020)

Efendioglu, I.H.: Metaverse concepts and marketing. In: Handbook of Research on Consumer Behavioral Analytics in Metaverse and the Adoption of a Virtual World, pp. 224–252 (2023)

Garda, B.: The journey of the tourism industry from digital marketing to metaverse network. In: Economic and Social Implications of Information and Communication Technologies, pp. 134–150. IGI Global (2023)

Gursoy, D., Malodia, S., Dhir, A.: The metaverse in the hospitality and tourism industry: an overview of current trends and future research directions. J. Hosp. Market. Manag. **31**(5), 527–534 (2022)

Hadi, A.A., Alnoor, A., Abdullah, H.O.: Socio-technical approach, decision-making environment, and sustainable performance: role of ERP systems. Interdiscip. J. Inf. Knowl. Manag. **13**, 397–415 (2018)

Hair Jr, F., Joe, M.S., Hopkins, L., Kuppelwieser, V.G.: Partial least squares structural equation modeling (PLS-SEM) an emerging tool in business research. Eur. Bus. Rev. **26**(2), 106–121 (2014)

Hennig-Thurau, T., Ognibeni, B.: Metaverse marketing. NIM Market. Intell. Rev. **14**(2), 43–47 (2022)

Hien, N.N., Nhu, T.N.H.: The effect of digital marketing transformation trends on consumers' purchase intention in B2B businesses: the moderating role of brand awareness. Cogent Bus. Manag. **9**(1), 2105285 (2022)
Hollensen, S., Kotler, P., Opresnik, M.O.: Metaverse–the new marketing universe. J. Bus. Strategy (2022)
Indumathi, R.: Influence of digital marketing on brand building. Int. J. Mech. Eng. Technol. (IJMET) **9**(7), 235–243 (2018)
Khwaja, M.G., Mahmood, S., Zaman, U.: Examining the effects of eWOM, trust inclination, and information adoption on purchase intentions in an accelerated digital marketing context. Information **11**(10), 478 (2020)
Kye, B., Han, N., Kim, E., Park, Y., Jo, S.: Educational applications of metaverse: possibilities and limitations. J. Educ. Eval. Health Prof. **18** (2021)
Lamba, S.S., Malik, R.: Into the metaverse: marketing to Gen Z consumers. In: Applying Metalytics to Measure Customer Experience in the Metaverse, pp. 92–98. IGI Global (2022)
Lee, J.Y.: A study on metaverse hype for sustainable growth. Int. J. Adv. Smart Convergence **10**(3), 72–80 (2021)
Lim, W.Y.B., et al.: Realizing the metaverse with edge intelligence: a match made in heaven. IEEE Wirel. Commun. (2022)
Melović, B., Jocović, M., Dabić, M., Vulić, T.B., Dudic, B.: The impact of digital transformation and digital marketing on brand promotion, positioning, and electronic business in Montenegro. Technol. Soc. **63**, 101425 (2020)
Morris, N.: Understanding digital marketing: marketing strategies for engaging the digital generation. J. Direct, Data Digit. Mark. Pract. **10**(4), 384–387 (2009)
Mystakidis, S.: Metaverse. Encyclopedia **2**(1), 486–497 (2022)
Nalbant, K.G., Aydin, S.: Development and transformation in digital marketing and branding with artificial intelligence and digital technologies dynamics in the Metaverse universe. J. Metaverse **3**(1), 9–18 (2023)
Narin, N.G.: A content analysis of the metaverse articles. J. Metaverse **1**(1), 17–24 (2021)
Ning, H., et al.: A survey on the metaverse: the state-of-the-art, technologies, applications, and challenges. IEEE Internet Things J. (2023)
Omar, A.M., Atteya, N.: The impact of digital marketing on the consumer buying decision process in the Egyptian market. Int. J. Bus. Manag. **15**(7), 120–132 (2020)
Purwanto, A.: How the role of digital marketing and brand image on food product purchase decisions? An empirical study on Indonesian SMEs in the digital era. J. Ind. Eng. Manag. Res. (2022)
Rani, K., Singh, S.: Metaverse: an innovative platform for digital marketing. In: Cultural Marketing and Metaverse for Consumer Engagement, pp. 215–223. IGI Global (2023)
Rathore, B.: Metaverse marketing: novel challenges, opportunities, and strategic approaches. Eduzone Int. Peer Rev. Refereed Multidisc. J. **7**(2), 72–82 (2018)
Satria, R., Hasmawaty, A.R.: Pengaruh digital marketing dan brand awareness terhadap Penjualan Produk KartuAS Telkomsel Cabang Palembang. Jurnal Nasional Manajemen Pemasaran & SDM **2**(3), 160–171 (2021)
Setiawan, K.D., Anthony, A.: The essential factor of metaverse for business based on 7 layers of metaverse–a systematic literature review. In: 2022 International Conference on Information Management and Technology (ICIMTech), pp. 687–692. IEEE, August 2022
Shevlin, R.: JPMorgan opens a bank branch in the metaverse (2022). https://www.forbes.com/sites/ronshevlin/2022/02/16/jpmorgan-opens-a-bank-branch-in-the-metaverse-but-its-not-for-what-you-think-its-for/?sh=61377f5d158d
Sigar, E.T., Massie, J.D., Pandowo, M.H.: The influence of consumer behavior and digital marketing on purchase decision at Grabfood in Manado. Jurnal EMBA: Jurnal Riset Ekonomi, Manajemen, Bisnis dan Akuntansi **9**(4), 53–64 (2021)

Tan, G.W.H., et al.: Metaverse in marketing and logistics: the state of the art and the path forward. Asia Pac. J. Market. Logistics (2023)

Tlili, A., et al.: Is metaverse in education a blessing or a curse: a combined content and bibliometric analysis. Smart Learn. Environ. **9**(1), 1–31 (2022)

Wang, F.Y., Qin, R., Wang, X., Hu, B.: Metasocieties in metaverse: metaeconomics and meta management for meta enterprises and megacities. IEEE Trans. Comput. Soc. Syst. **9**(1), 2–7 (2022)

Wang, Y., et al.: A survey on metaverse: fundamentals, security, and privacy. IEEE Commun. Surv. Tutorials (2022)

Wong, I.A., Lin, S.K., Lin, Z.C., Xiong, X.: Welcome to stay-at-home travel and virtual attention restoration. J. Hosp. Tour. Manag. **51**, 207–217 (2022)

Yadav, M., Rahman, Z.: Measuring consumer perception of social media marketing activities in e-commerce industry: scale development & validation. Telematics Inform. **34**(7), 1294–1307 (2017)

Yim, M.Y.C., Chu, S.C., Sauer, P.L.: Is augmented reality technology an effective tool for e-commerce? An interactivity and vividness perspective. J. Interact. Mark. **39**(1), 89–103 (2017)

Zhang, D., Chadwick, S., Liu, L.: The Metaverse: Opportunities and Challenges for Marketing in Web3 (2022). SSRN 4278498

Zhao, Y., et al.: Metaverse: perspectives from graphics, interactions, and visualization. Vis. Inf. **6**(1), 56–67 (2022)

The Effect of Religion on Metaverse Marketing

Bakhtiar Aubaid Sharif(✉)

Technical Business Administration Department, Technical College of Administration, Sulaimani Polytechnic University, Sulaimania, Kurdistan Region, Iraq
bakhtiar.sharif@spu.edu.iq

Abstract. The objective of the current study is to clarify and quantify how religion affects metaverse marketing among employees at Sulaimani Polytechnic University's Technical College of Engineering. The problem that served as the impetus for the current investigation was articulated by a theoretical and practical query. Since religion is one of the key issues in the college environment, we investigated the effect of religion on metaverse marketing. In this current work, the researcher employed the analytical descriptive method. The researcher used a hypothetical model that depicts the nature of the interaction and influence between the think about factors in arrange to achieve the study's objectives and give an answer to the earlier inquiry. The majority of the study's participants are from Sulaimani Polytechnic University's Technical College of Engineering. The questionnaire served as the main instrument for gathering data and information for the study. The employees were given 130 questionnaires, and 111 of them were returned for study. Based on the findings, the study came to the conclusion that religion encompasses a coordinate impact on metaverse marketing and that there is a reciprocal relationship between the two variables.

Keywords: Religion · Marketing · Islamic marketing · Metaverse · Religious groups

1 Introduction

The basic objective of businesses is to maximize shareholder wealth and increase returns on operations. Business organizations exist to generate profit. In addition, high sales rates result in high rates of return. It is important to note that the ratio of sales depends on consumer behavior; however, consumer behavior is influenced by religion because it affects people's values and actions (Manhal et al., 2013). As a result, in many societies, religion plays a significant role in the lives of individuals. It shapes the way they think, make decisions, and behave, what to eat and what not to eat, attitude, and behavior that this has on the consumer's purchasing decisions for particular goods that may not be permitted in some religions (Barrera and Shah, 2023). Since the success of a company depends on the marketing of its products and services, every business must now consider the religion of the region in which it operates (Atiyah, 2022). Additionally, in order to increase the effectiveness of their advertising, international advertisers and advertising agency managers must have a thorough understanding of how different religious beliefs and their

M. Al-Emran et al. (Eds.): IMDC-IST 2024, LNNS 876, pp. 129–143, 2023.
https://doi.org/10.1007/978-3-031-51300-8_9

strength affect offense toward the advertising of contentious products. By appropriately incorporating religious sentiment, religion can be advantageously employed in marketing campaigns (Buhalis et al., 2023). Customers have a stronger emotional connection with the brand or product and begin to like it. Because each religious culture offers slightly different incentives for human behavior, the relative strength of a religion's positive and negative institutional effects on global trade varies from one religious culture to another (Hollensen et al., 2023). Many businesses today employ workers from various countries and produce and sell goods globally (Mansori, 2012; Serwah et al., 2022; Alnoor et al., 2022).

In organizational research and management practice, religion has typically been neglected or even outlawed. This is a stunning omission from the letters on business, society, and corporate ethics (Abbas et al., 2023; Chew et al., 2023). As a source of moral standards and beliefs, religion has historically played a significant role in the vast majority of societies and continues to be relevant in almost every society today. Furthermore, regional, national, or indigenous cultures—which, in many parts of the world, are greatly influenced by religious belief systems and religious institutions—determine expectations for ethical business conduct. Therefore, the impact of religion on marketing has not yet been extensively researched. The reason of this ponder is to fill within the information hole that right now exists with respect to the impact of religion and how it impacts people's decisions to purchase necessities of life based on product marketing, as well as to demonstrate the right method of advertising while taking into account product or service marketing. Additionally, managers of international marketing and advertising agencies need to understand how different religious beliefs and how they negatively affect outrage toward the marketing of goods and services in order to develop effective advertising without offending or alienating their target audience (Ab Talib, Ai Chin, & Fischer, 2017).

A significant amount of people's bases for purchasing goods and services around the world depend on their religious beliefs and perspectives. The following are some of the implications of this research:

1. Religion has a significant role in marketing campaigns, and campaigns should be properly created to ensure a business's success.
2. Since consumers have an emotional connection to religion, marketing initiatives can benefit from this by appropriately incorporating religious emotions.
3. The religion of the area must be considered when advertising for a product because, according to Pew Forum, 80% of people worldwide identify as members of a particular religion in 2012. However, the content and the marketing style or design must be appropriate for the locals, so faith-based marketing acknowledges the significance of religion to customers and makes an effort to connect the product being sold to religion.

2 Literature Review

Kuzma et al. (2009) investigated how mega-churches affect marketing strategy by examining the effects of religion on various marketing disciplines and business in general. The study discovered that religion, which is regarded as one of the psychological aspects of consumers, has successfully influenced the method in which they choose their products. The study also found that consumers' religious beliefs influence their decision to buy. In a similar vein, another article (Fam et al., 2004; Mohammed et al., 2022) has examined the influence of religion and belief intensity on behavior towards marketing of a four different products and services, namely, (gender/sex related products, social/political groups, health and care products, and addictive products), through conducting a questionnaire on 1393 people, distributed on four religious groups, namely (Buddhism, Christianity, Islam, and non-religious believers) in six different countries. The multivariate analysis of variance test was used in the study to demonstrate statistically significant differences in outcomes between the groups. However, the findings of the study "Religious beliefs and advertising of offensive products" revealed that, in comparison to the other three religions, Islamic people found the advertising of gender/sex-related products, social/political groups, and health and care products to be the most disrespectful. The study also revealed that religion is a long-lasting phenomenon that warrants further study. Based on Migdalis et al., (2014) work, the methodology used in this paper was quantitative data with a questionnaire survey that was distributed over people with different nationality, namely Greece (Athens, n = 100), Bulgaria (Sofia, n = 80), United Arab Emirates (UAE) (Dubai, n = 70, Jeddah, n = 30), and people living in UAE or Saudi Arabia but were foreigners (n = 44). Another study attempted to investigate the relationship between religion and its impact on consumption behavior. And religion. The analyses included both descriptive and statistical tests to shed further light on the relationships and the statistically significant mean value differences. The sample consisted of 330 surveys, of which 324 had responses (98.1% response rate). The results of this study revealed that although people from Greece and Bulgaria's buying decisions are impacted by their religion by 8% and 5%, respectively, those from Saudi Arabia's shopping decisions are affected by religion by 54% and 54%, respectively. In addition, based on their religious preferences, Christian and Muslim consumers exhibit a variety of behaviors. It is suggested that religious micro cultures should be seen as unique consumer segments, which can call for alternative marketing initiatives.

According to Muchilwa et al. (2011) the pharmaceutical sector caters to people of diverse ethnic backgrounds and religious beliefs. A survey was used as the paper's technique. The questionnaires, which were distributed across 137 local and foreign pharmaceutical companies in Kenya and used a cross-sectional descriptive survey design, required the manager or director of the marketing department at each company to respond. Both qualitative and quantitative data were collected. The researcher came to the conclusion from this paper that all respondents were in agreement that religion is a factor that affects organizational structures and operations, and that 94% of respondents agreed that this is true in their own companies. In addition, the paper provides evidence that the department of the company most affected by religion is marketing, and that the most affected marketing strategies by religion are place-based.

(Rao, 2012) used a sample of 9 companies with multinationality employees, 7 of whom were senior executives in U.S. multinational firms in India and 2 Indian companies, to examine the existence of religions in Indian workplaces through cultural values, beliefs, and management practices and their influence on international human resources management practices in managing diversity and the other seven were in technology sector. The data was gathered using an interviewing style, and questions were asked about the influence of religion on management styles and philosophy, HR practices like recruitment, selection, and retention, women at work, attitude toward work, observances, holidays, appearance or behavior at work, and finally religious conflicts at work. The findings revealed that every interviewee (100% of the sample) acknowledged that religion had an impact on the workplace. The respondents said that knowing about Indian religions helped them better comprehend the subordinates they handled. All interviewees added that they did not base any HR decisions on religion. In addition, every interviewee believed that religion might lead to significant conflict at work. Additionally, each interviewee was able to provide concrete examples to demonstrate the influence of religion on a surface level. They acknowledged that they observed numerous religious holidays at work.

2.1 Religion

When we accept something as true despite uncertainty or our inability to verify it, this is known as belief. Everybody has a unique set of beliefs about themselves and the world they inhabit. Mutually reinforcing beliefs create belief systems, which can be religious, philosophical, or ideological (Sheth and Mittal, 2004; AL-Fatlawey et al., 2021). Religions are philosophical systems that connect people with the divine. Religion is a collection of cultural frameworks, worldviews, and philosophical systems that connects people to spirituality and, occasionally, moral ideals. Numerous religions have stories, symbols, rituals, and sacred histories that are intended to explain the origins of life or the universe or to give life a purpose. Their theories about the universe and human nature frequently serve as the foundation for morality, ethics, religious laws, or a desired way of life. Numerous religions have clergy, rules defining what constitutes loyalty or membership, laity congregations, regular worship services or gatherings, holy sites (natural or man-made), and/or scriptures. Religious activities include preaching, remembering the deity or deities, offering sacrifices, festivals, feasts, trance, initiations, funeral rites, weddings, meditation, music, painting, dance, and other aspects of human culture. However, some or all of these aspects of organization, belief, or practice are absent from some religions (Usinier and Stolz, 2014; Gatea and Marina, 2016).

People have believed in the spiritual side of existence from the beginning of time. Whether it was sun worship, god and goddess worship, knowledge of good and evil, or holy knowledge, many human societies have left historical traces of their belief systems. Stonehenge, the Bamiyan Buddhas, the Almudena Cathedral in Madrid, Uluru in Alice Springs, the Bahá' Gardens in Haifa, Japan's sacred mountain Fujiyama, the Kaaba in Saudi Arabia, and the Golden Temple in Amritsar are all examples of structures that attest to the spiritual experience of humans, which may be a subjective reality or the outcome of our search for an explanation of the purpose of life and our place in the universe (Santoro and Siliman, 2015; Sak et al., 2023).

2.2 Religion in Marketing

Business and religion are increasingly interacting. On the one hand, religious organizations employ sophisticated marketing techniques to recruit new members. The resources available include members, volunteers, money, and public support, to name just a few.

On Contrarily, the commercial sector appropriates information with a religious or spiritual theme. Promote and market the products and services of the business. It is possible to look at this complex composition from both angles, which presents a number of significant difficulties. Research on the subject of the connection between religion and marketing suggests that society is undergoing a profound, ongoing transition. As seen by the academic work that has been done, an increasing number of academics are studying this link. Has been developed during the past 20 years, and in the last few years, it has seen substantial expansion (Nardella, 2014). The objective for the scholars was to characterize the relationship's essential elements, appreciate its characteristics, and place it within broader historical and social settings. This generated a dialogue that turned up signs of understanding between the marketing and religious sectors. Although there are many areas of interaction and overlap, it was once believed that these two universes belonged to two distinct worlds, or at least theoretically opposed cultural realms (Tariq, M., & Khan, M. A. 2017; Hadi et al., 2019).

2.3 Islamic Marketing

Understanding the connection between Islam and consumption and marketing strategies has drawn more attention in recent years. This interest is reflected in the creation of specialized journals, the growth in the quantity of research articles published, the planning of academic conferences and executive workshops, and the creation of reports from well-known consultants. This immediate and intense focus is closely related to the question of why. Why is marketing and Islam such a hot topic right now, and why is that? (Sahlaoui and Bouslama, 2016).

A closer look at the literature suggests that the motivation behind this interest is the perception of Muslims as an underserved and promising consumer segment. The rise in Muslim consumer exposure is closely related to their purchasing power, just like it is with other non-mainstream consumer groups in the US like Blacks and Hispanics. This strength is particularly seen in the emergence of a middle class among Muslims who, despite living in different parts of the world, share a common desire for consumption and financial means to buy name-brand products (Ishak & Abdullah, 2012). A new breed of enterprises that are fervently pursuing Islamic beliefs as well as economic goals in both Muslim-majority and Muslim-minority countries are becoming more prominent alongside the Muslim middle class (Mansori and Shaheen, 2012). Overall, it seems that the economy and shifting demography are having an effect. Muslim consumers now have more purchasing power, and Muslim entrepreneurs are doing better. Islamic marketing is starting to gain scholarly recognition and management appeal. The phrase "Islamic marketing," on the other hand, raises considerable concern.

The "Islamic" accent has the power to exacerbate divisions rather than promote communication. It first proposes that in order to advertise to Muslim customers, the marketing strategy must have an Islamic feel. It is believed that this form of marketing is different from conventional marketing. Second, it implies that Islamic marketing targets Muslim consumers, who differ from other consumers, and that it employs tools, skills, and strategies that are pertinent to and appealing to this group. Last but not least, it is presumable that there is an already-existing and uniform Muslim customer category that marketers can target, reach, and somewhat predict. Such presumptions run the potential of giving rise to an essentialist perspective, which produces a somewhat rigid and cliched outcome. Muslim consumers, businesses, and associated marketing and consumption patterns are all widely known (Sandikci, 2011).

2.4 Market Share

A company's market share is determined by what much of a customer's overall purchases of a good or service goes to that business. In other words, if the general public buys 100 soaps and one corporation buys 40 of them, that company has a 40% market share (Brodrechtova, 2008). Market share can take many different forms and dimensions. Market shares may be calculated based on volume or value. Value market share is calculated using a company's overall percentage of total sector sales. Volume is the genuine number of units sold by a company as compared to the full number of units sold within the showcase. The relationship between esteem and volume showcase offers is once in a while straight; for case, a unit may have a tall esteem but moo numbers, which shows a tall esteem showcase share but a moo volume advertise share.

Value market share is calculated using a company's overall percentage of total sector sales. Volumes are the proportion of a company's sales from the total market sales of all units. The relationship between value and volume market shares is rarely linear; for example, a unit may have a tall esteem but moo numbers, which demonstrates a tall esteem advertise share but a moo volume showcase share. In businesses like FMCG, where items are moo esteem, huge volume, and there are numerous freebies, comparing esteem advertise share is common. The noteworthiness of advertise share: Showcase share could be a gage of buyer preference for a item over competing merchandise. A better boundary to section for modern competitors and higher deals regularly take after from having a bigger showcase share. A pioneer will pick up more from showcase development in the event that they have a greater advertise share than the competition. In arrange to maintain its possess development, a advertise pioneer as measured by showcase share must moreover grow the advertise (Sharif et al., 2020). Thus, we assume that:

H1: There is an effect of religion on metaverse marketing.

2.5 Metaverse and Marketing

The world of Metaverse is a virtual world that simulates the real world. Thus, it relies substantially on virtual reality technologies, which are formerly wide in numerous countries around the world. Experts anticipate that the number of its druggies will reach 58.9 million druggies in the United States alone. Given the enormity of the world of Metaverse, it's anticipated to be the successor to the Internet that we're presently passing. With technologies similar as 5G networks, artificial intelligence, and stoked reality, Metaverses are anticipated to grow fleetly (Mystakidis, 2022).

Marketing is the practice of promoting and dealing a company's products or services. It includes the four rudiments of marketing price, product, creation and place. Metaverse marketing and advertising can give new places to buy and new ways to promote. The elaboration of marketing keeps pace with changes on the web. When Web1.0 began, the thing of marketing was to have a website with a business's contact information. Also

Web2.0 started connecting people and collecting their hunt history to make the stoner experience more individualized. Now, with Web3.0 comes a happier experience. According to studies, there are 400 million unique and active druggies penetrating Metaverse every month. To reach these druggies, companies need to follow generational cult who watch about the metaverse.

Metaverse is new, so companies can look innovative to consumers since not everyone is using Metaverse yet. Traditional marketing has further competition, so Metaverse may be a way for companies to stand out and produce their own way of advertising (Ball, 2021; Albahri et al., 2023). Because it's new, there are further pitfalls in Metaverse than in traditional advertising because results can be delicate to measure and not all consumers use it. On the other hand, one of the most intriguing aspects of Metaverse is decentralization. Unlike Facebook where associations and individualities use a platform possessed by another company- they've the occasion to produce their own world and produce the terrain they see. Traditional advertising has platforms that review all content before going live including online spots similar as YouTube, Facebook and Instagram and other advertising media – similar as pamphlets, radio, billboards and banner advertisements (Buhalis, 2020).

3 Methodology

Quantitative research was used to adopt a systematic approach to investigating the topic of the research question, examining the gap in its cognitive and commonsense measurements, and analyzing and translating it in accordance with the logical strategy, which is reflected in the research objectives and the outcome of the various qualities and numerous methods of measuring the subject of the research. Quantitative research, which receives the information entrance within the overview of the wonders shaped by the study issue, may be a strategy branded by a thorough representation of the precise facts and data acquired in order to draw conclusions and offer recommendations. The study population serves as the foundation for establishing the research variables that define its explicit objectives. Therefore, it is essential to precisely define this population. The employees of Sulaimani Polytechnic University's Technical College of Engineering were chosen as the study's topic. The study population was decided upon as being the staff at the Technical College of Engineering.

Choosing the study sample's size is a crucial and challenging phase in the research design process. The quality and accuracy of study findings are significantly impacted by inadequate and large sample sizes. As a result, one of the crucial steps in the phases of the research design process is to decide on the size of the sample that is suitable and acceptable for the study. Results from the tiny sample size are neither reliable or accurate. On the other hand, an extremely large sample size wastes the researcher's time and effort. Here, as shown in Table 1, it was necessary to choose the employees of the Technical College of Engineering using a randomly selected sample. 150 employees of Sulaimani Polytechnic University's Technical College of Engineering made up the study's population. The study's sample size was determined using the sampling formula developed by Krejcie and Morgan (1970). The population sample size was thus determined to be (108). Only (111) of the (130) respondents from the Technical College of

Engineering at Sulaimani Polytechnic University returned the survey questionnaire, as shown in Table 1.

Table 1. Employees in the Technical College of Engineering at Sulaimani Polytechnic University.

No	Name of University	College	No. of Employees	Distributed forms	Returned forms	Distribution ratio %
1	Sulaimani Polytechnic University	Technical College of Engineering	150	130	111	85.38%

4 Results

According to Zikmund (2003), the degree of reliability refers to how error-free a measure is and how consistently reliable its results are. Additionally, a reliability test was performed to determine whether an instrument is stable and consistent in measuring a concept to the point where the error in the observed score is minimal. According to Hoe (2008), Cronbach's alpha is most useful for indicating scale reliability in terms of item equivalence within single-construct scales. Cronbach's alpha can be regarded as a totally appropriate indicator of the intermediate consistency reliability, claim (Sekaran & Bougie, 2016).

Table 2. Reliability Test.

Number	Variables	Number of class	Value (α)
1	Religion	7	0.943
2	Metaverse Marketing	7	0.909
Total	Total	14	0.921

The reliability of the participations was determined using alpha Cronbach, as shown in Table 2. The Alpha Cronbach coefficient, on the other hand, was employed to assess the accuracy of the responses provided by the research sample's participants and to guarantee the stability of the scale in use. According to the outcomes of the computer analysis, it was determined that the value of the Alpha Cronbach coefficient at the combined level of the two variables (marketing and religion) is equal to (0.921). As a result, the alpha Cronbach score is (0.921), and this indicates that the questionnaire is extremely reliable. According to Kerlinger and Lee (2000), coding is a method used to make it clear how to translate respondent data and question answers to particular categories for the analysis procedures. As a result, Table 3 shows how variables, dimensions, and items are coded.

The researcher used data from a sample of respondents' demographic respondents. To learn more about each respondent who took part in the survey, the demographic

Table 3. Coding of Variables and Items.

Variables	Variable Code	Numbers of Items
Religion	R	7
Metaverse Marketing	M	7

information of the respondents was obtained. Age, gender, marital status, degree of education, and religion were all questions that respondents had to answer. Instead of asking for specific information, the questions were made such that respondents could select their responses from a list of categories. Using SPSS version 24, Table 4 displays the demographic profile of the respondents.

Table 4. Profile of the Respondents' Demographic Factors.

Details		Frequency	Percent	Valid Percent	Cumulative Percent
Age	Under 25 years	96	86.5	86.5	86.5
	25–35 years	11	9.9	9.9	96.4
	36–45 years	4	3.6	3.6	100
	More than 45 years	0	0.0	0.0	100
	Total	111	100.0	100.0	
Gender	Female	59	53.2	53.2	53.2
	Male	52	46.8	46.8	100
	Total	111	100	100	
Marital Status	Single	12	10.8	10.8	10.8
	Married	99	89.2	89.2	100
	Total	111	100	100	
Educational Level	High School	7	6.3	6.3	6.3
	Diploma	7	6.3	6.3	12.6
	Bachelor	95	85.6	85.6	98.2
	Postgraduate	2	1.8	1.8	100
	Total	111	100	100	

(*continued*)

Table 4. (*continued*)

Details		Frequency	Percent	Valid Percent	Cumulative Percent
Religion	Muslim	105	94.6	94.6	94.6
	Christian	2	1.8	1.8	96.4
	Kakaye	0	0.0	0.0	96.4
	Other	4	3.6	3.6	100
	Total	111	100	100	

Data from Table 5 repetition distributions (mean, standard deviation, coefficient of variance, and relative importance) show that (Religion) is a key factor in the explanatory factors. The mean value of this variable is 3.92. Additionally, there was a 1.19 standard deviation and a 78.48% relative importance. The percentage of respondents who chose "strongly agree" to (42.08%), "agree" by (29.09%), "agree" slightly (13.9%), and "disagree" to (9.01%), "strongly disagree" (5.92%).

Table 5. Description of variables.

Questions	Strongly Disagree NO %	Disagree	Not Sure NO %	agreeNO %	strongly agree NO %	Mean NO %	S.D NO %
X_1	10	12	14	30	45	3.79	1.32
	9.01	10.81	12.61	27.03	40.54		
X_2	5	20	12	40	34	3.70	1.21
	4.50	18.02	10.81	36.04	30.63		
X_3	8	12	22	14	55	3.86	1.33
	7.21	10.81	19.82	12.61	49.55		
X_4	4	10	21	33	43	3.91	1.12
	3.60	9.01	18.92	29.73	38.74		
X_5	8	6	15	44	38	3.88	1.16
	7.21	5.41	13.51	39.64	34.23		
X_6	7	3	12	32	57	4.16	1.13
	6.31	2.70	10.81	28.83	51.35		
X_7	4	7	12	33	55	4.15	1.08
	3.60	6.31	10.81	29.73	49.55		
Sum	46	70	108	226	327	3.92	1.19
	5.92	9.01	13.90	29.09	42.08		

The statistics in Table 6 on repetition distributions (mean, standard deviation, coefficient of variance, and relative importance) show that (Metaverse Marketing) is the main area of attention for the explanatory variables. The mean value of this variable is 3.89. Additionally, there was a 1.17 standard deviation and a 77.86% relative importance. The percentage of respondents who chose "strongly agree" was 40.15 percent, followed by "agree" responses from 29.6 percent, "agree" responses from 15.83 percent, and "disagree" responses from 8.14% and 6.18%, respectively.

Table 6. Description of variables.

Questions	Strongly Disagree NO %	Disagree NO %	No NO %Sure	Agree NO %	strongly agree NO %	Mean NO %	S.D NO %
X_1	5	16	10	33	47	3.91	1.23
	4.50	14.41	9.01	29.73	42.34		
X_2	8	12	21	30	40	3.74	1.26
	7.21	10.81	18.92	27.03	36.04		
X_3	4	8	16	23	60	4.14	1.13
	3.60	7.21	14.41	20.72	54.05		
X_4	2	5	18	33	53	4.17	0.98
	1.80	4.50	16.22	29.73	47.75		
X_5	2	5	30	44	30	3.86	0.93
	1.80	4.50	27.03	39.64	27.03		
X_6	20	10	14	30	37	3.49	1.48
	18.02	9.01	12.61	27.03	33.33		
X_7	7	8	14	37	45	3.95	1.18
	6.31	7.21	12.61	33.33	40.54		
Sum	48	64	123	230	312	3.89	1.17
	6.18	8.24	15.83	29.60	40.15		

Table 7 shows a substantial positive statistical correlation (0.855) between the dependent and independent variables, with a significance value of 0.000 and a lower bound of 0.05. This suggests that the second hypothesis is correct because there is a strong association between marketing and religion Table 8.

The expected variable value (metaverse marketing) is significant for the value of (p-value) of (0.000), which indicates that it is less than the significant level of 0.05, which means that it is statistical significance and accept the alternative hypothesis, as can be seen from the table we created below. It is evident that the independent variable factor (religion) has a value of (0.981) while the coefficient of determination has a value of (0.731), the total variance is equal to (73.1%), and the other variables (26.1%) are the result of random error.

Table 7. Relationship between variables.

Dependent variable			Independent variable
Metaverse Marketing			
Sig	Sample	Correlation	
0.000	111	0.855	Religion
The level of significance at level 0.05 There is a relationship between the statistical function between independent variable and dependent variable			

Table 8. Regression analysis.

Model	Unstandardized Coefficients		T Test	Sig	Model summary		ANOVA table	
					R	R^2	F Test	Sig
	B	Std. Error						
Constant	0.454	0.145	4.544	0.000	0.855	0.731	461.358	0.000
Religion	0.981	0.023	12.5433	0.000				

5 Discussion

The results showed the consumer's interest in purchasing the product whose cover contains a halal symbol, according to his/her different religious and social beliefs affecting that. We found that there is a direct effect of religion on metaverse marketing, the final results of the field study 150 results previously obtained from Statistical analysis of a questionnaire distributed to the respondents in the study community, it became clear to us that despite the differences in consumers in terms of (age, gender, marital status, educational level, religion), they confirm that religion has a direct effect on metaverse marketing, and this indicates that our hypothesis is correct. Thus, the consumer is interested in providing commodities commensurate with his/her beliefs, as this is limited to the use of religious references on the packaging of the product. Even if these signs are not present, the credibility of the product present in the producing country affects highly positively on purchase decisions and consumer behavior. This is shown by the study of (Kuzma et al., 2009), which investigated how major churches influence marketing strategy by examining the effects of religion on various marketing disciplines and business in general. The results of this research are supported by another article (Fam et al., 2004) on the effect of religion and belief on behavior towards the marketing of four different products and services. Previous studies such as (Rao, 2012) used a sample of 9 companies with multinational employees, 7 of whom were senior executives of US multinational companies in India and 2 Indian companies, to examine the presence of religions in Indian workplaces through cultural values, beliefs, managerial practices and their impact on resource management practices International Human in Diversity

Management while the other seven were in the technology sector. In comparison with the results of previous research, the results of this study support the results of previous studies and confirm the validity of our hypothesis.

6 Conclusions

Several findings were made using the findings of earlier research and the findings of the field study, which can be summed up as follows:

1. Religion addresses the most fundamental part of the person's cognitive universe.
2. How an organization implements its operations and services has an impact on the marketing process.
3. Shopping is viewed in the world differently depending on the religion. Each religion approaches marketing from a distinctively religious vantage point.
4. The goal is distinct from the marketing model and theology. Religion attempts to establish a genuine interest for society in accordance with the framework of religion, while marketing seeks to accomplish the highest number of returns and profits.
5. While it is true that religion can influence how people behave generally and how they use things specifically, the usefulness of the pious subculture as a tool for dividing customers is mostly untapped.
6. The researcher came to some intriguing conclusions about the influence of pious ideals on consumer behavior.

7 Recommendations

1. The requirement that the person promoting or marketing goods and commodities be distinguished by the virtues of religion and morals.
2. Before purchasing something, it must be subject to social and religious regulations.
3. Offering seminars and training sessions on religious shopping to staff members of marketing organizations.
4. Expanding study and studies of the relationship between marketing and religion.
5. The requirement to stay current with trends and modifications in the international market.
6. The importance of focusing on the human cadres who work in the institutions.

The necessity to develop and work to improve the goods and services offered to customers.

References

Ab Talib, M.S., Ai Chin, T., Fischer, J.: Linking Halal food certification and business performance. British Food J. **119**(7), 1606–1618 (2017)

Abbas, S., et al.: Antecedents of trustworthiness of social commerce platforms: a case of rural communities using multi group SEM & MCDM methods. Electron. Comm. Res. Appl. 101322 (2023)

Albahri, O.S., et al.: Corrigendum to "Novel dynamic fuzzy Decision-Making framework for COVID-19 vaccine dose recipients. J. Adv. Res. 37 (2022) 147–168]. J. Adv. Res. **45**, 193 (2023)

AL-Fatlawey, M.H., Brias, A.K., Atiyah, A.G.: The role of strategic behavior in achievement the organizational excellence" Analytical research of the manager's views of Ur State Company at Thi-Qar Governorate". J. Adm. Econom. **10** (37) (2021).

Alnoor, A., et al.: How positive and negative electronic word of mouth (eWOM) affects customers' intention to use social commerce? A dual-stage multi group-SEM and ANN analysis. Int. J. Human–Comput. Interact. 1–30 (2022).

Atiyah, A.G.: Effect of temporal and spatial myopia on managerial performance. J. La Bisecoman **3**(4), 140–150 (2022)

Ball, M.: Framework for the Metaverse—MatthewBall.vc', Retrieved 25 August 2022, from https://www.matthewball.vc/all/forwardtothemetaverseprimer

Barrera, K.G., Shah, D.: Marketing in the metaverse: conceptual understanding, framework, and research agenda. J. Bus. Res. **155**, 113420 (2023)

Brodrechtova, Y.: Determinants of export marketing strategies of forestproducts companies in the context of transition. The case Slovakia, Forest Policy Econom. **10**(7), 450–459 (2008)

Buhalis, D.: Technology in tourism-from information communication technologies to eTourism and smart tourism towards ambient intelligence tourism. Persp. Article, Tourism Rev. **75**(1), 267–272 (2020)

Buhalis, D., Leung, D., Lin, M.: Metaverse as a disruptive technology revolutionising tourism management and marketing. Tour. Manage. **97**, 104724 (2023)

Calantone, R.J., Kim, D., Schmidt, J.B., Cavusgil, S.T.: The influence of internal and external firm factors on international product adaptation strategy and export performance. J. Bus. Res. **59**(2), 176–185 (2006)

Chew, X., Khaw, K.W., Alnoor, A., Ferasso, M., Al Halbusi, H., Muhsen, Y.R.: Circular economy of medical waste: novel intelligent medical waste management framework based on extension linear Diophantine fuzzy FDOSM and neural network approach. Environ. Sci. Pollut. Res., 1–27 (2023). https://doi.org/10.1007/s11356-023-26677-z

Fam, K.S., Waller, D.S., Erdogan, B.Z.: 'The influence of religion on attitudes towards the advertising of controversial products', European Journal of marketing. Gravetter, F.J., Wallnau, L.B (2004)

Gatea, A.A., Marina, V.: Higher education funding in Iraq in terms of the experience of particular developed countries. Int. J. Adv. Stud. **6**(1), 8–17 (2016)

Hadi, A.A., Alnoor, A., Abdullah, H., Eneizan, B.: How does socio-technical approach influence sustainability? Considering the roles of decision making environment. Appl. Decision Sci. Bus. Manage. 55 (2019)

Hoe, S.L.: Issues and procedure in adopting structural equation modeling technique. J. Appl. Quant. Method **31**, 76–83 (2008)

Hollensen, S., Kotler, P., Opresnik, M.O.: Metaverse–the new marketing universe. J. Bus. Strat. (ahead-of-print) (2022).

Kerlinger, F.N., Lee, H.B.: 'Foundations of behavioral research (4th ed.)', Holt, NY: Harcourt College Publishers (2000)

Kuzma, A., Kuzma, A., Kuzma, J.: How religion has embraced marketing and the implications for business. J. Manage. Market. Res. **2**, 1 (2009)

Manhal, M., Al-khalidi, A., Hamad, Z.: Strategic network: managerial myopia point of view. Manage. Sci. Lett. **13**(3), 211–218 (2023)

Mansori, S.: 'Impact of religion affiliation and religiosity on consumer innovativeness. World Appl. Sci. J. **7**(3). 301–307 (2012). ISSN 1818–4952

Migdalis, N.I., Tomlekova, N., Serdaris, P.K., Yordanov, Y.G.: The impact of religion on shopping behavior. Int. J. Manage. Res. Rev. **4**(12), 1120 (2014)

Mohammed, M., Muhammad, A., Alnoor, A., Al-Maatoq, M.: The effect of adopting technology in intellectual capital. In: Proceedings of 2nd International Multi-Disciplinary Conference Theme: Integrated Sciences and Technologies, IMDC-IST 2021, 7–9 September 2021, Sakarya, Turkey (2022).

Muchilwa, R.C.: 'Influence of Religion on Marketing Strategies of International Pharmaceutical Firms. In: Kenya', Doctoral dissertation, University of Nairobi (2011)

Mystakidis, S.: Metaverse. Encyclopedia' **2**, 486–497 (2022). https://doi.org/10.3390/encyclopedia2010031

Nardella, C.: Studying religion and marketing, An introduction Sociologica, **8**(3), 0–0 (2014)

Rao, A.: Impact of religion in the Indian workplace. J. World Bus. **47**(2), 232–239 (2012)

Sahlaoui, M., Bouslama, N.: The marketing and islamic points of view. Am. J. Ind. Bus. Manag. **6**, 444–454 (2016)

Sak, M., Alnoor, A., Valeri, M., Bayram, G.E.: The role of digital transformation on women empowerment for rural areas: the case of Turkey. In: Tourism Innovation in the Digital Era (pp. 91–105). Emerald Publishing Limited (2023).

Sandıkçı, Ö: 'Researching Islamic marketing: past and future perspectives', Journal of Islamic marketing (2011)

Santoro, A., Siliman, D.: Religion and the Marketplace in the United States. Oxford University Press, Oxford (2015)

Sekaran, U., Bougie, R.: 'Research methods for business: A skill building approach', john wiley & sons

Serwah, A.M.A., KHAW, K.W., Yeng, C.S.P., Alnoor, A.: Customer analytics for online retailers using weighted k-means and RFM analysis. Data Analytics and Applied Mathematics (DAAM), pp. 1–7 (2023).

Sharif, B.A., Kaka Mala, H.M., Ali, N.A.: The relation between university education management and the vocational development. J. Garmian Univ. **6**(4), 312–329 (2020)

Sheth, J.N., Mittal, B.: Customer Behavior: A Managerial Perspective, 2nd edn. South-Western, USA (2004)

Tariq, M., Khan, M.A.: Offensive advertising: a religion based Indian study. J. Islamic Market. **8**(4), 656–668 (2017). https://doi.org/10.1108/JIMA-07-2015-0051

Usinier, J.C., Stolz, J.: Religions as Brands: New Perspectives on the Marketization of Religion and Spirituality. Ashgate Publishing Limited, Routledge (2014)

Zikmund, W.G., Babin, B.J., Carr, J.C., Griffin, M.: 'Business research methods 7th ed', Thomson/South-Western (2003)

Understanding Metaverse Adoption Strategy from Perspective of Social Presence and Support Theories: The Moderating Role of Privacy Risks

Abbas Gatea Atiyah[1], Mushtaq Alhasnawi[1,2](✉), and Muthana Faaeq Almasoodi[3]

[1] College of Administration and Economic, University of Thi-Qar, Nasiriyah, Iraq
{abbas-al-khalidi,alhasnawi78}@utq.edu.iq

[2] School of Management, Universiti Putra Malaysia, Seri Kembangan, Malaysia

[3] Department of Tourism Studies, College of Tourism Science, Kerbala University, Karbala, Iraq

Abstract. From time-to-time technological development offers technologies that respond to environmental requirements but are more complex. Metaverse is one of the most modern technologies that helps the individual to enter into a realistic, tangible experience, which is considered a revolution in the digital field. Entering the metaverse environment comes with concern about one's privacy. Such as the loss of personal data and the possibility of its transmission and use by others. In light of these obstacles, is it possible to utilize metaverse as an adopted strategy? This study aimed to test the possibility of adopting the metaverse strategy, by two main elements, which are social presence and social support. Taking into account privacy risks as an interactive variable. The study data were collected from (210) users and tested statistically by PLS-SEM approach to address the hypotheses of the study. A set of results was reached that showed the impact of social presence and social support on the adoption of metaverse. With the emergence of the impact of privacy risks in that relationship. The study concluded that adopting metaverse should take into account the privacy risks of users in dealing with digital technologies.

Keywords: Adoption Strategy of the Metaverse · Social Presence · Social Support · Privacy Risks · Metaverse Space · Physical and Virtual Existence

1 Introduction

In the rapidly developing digital age, metaverse technology has appeared as a tool that blurs the barriers between physical and virtual existence. With the evolving of this technology, it permeated our lives and revolutionized the field of interaction between individuals. Experimentation and exploration through the metaverse have increased greatly. Therefore, it has become necessary to understand this complex network and realize the factors that increase or decrease its adoption in the reality of business. Previous studies had a rarity in addressing integrate the factors that would raise and lower the possibility of adopting Metaverse in the business world (Mourtzis et al., 2022; Arpaci; Pellegrino et al., 2023; Chan et al., 2023). This is what is unique to our current study. The focus of

M. Al-Emran et al. (Eds.): IMDC-IST 2024, LNNS 876, pp. 144–158, 2023.
https://doi.org/10.1007/978-3-031-51300-8_10

many previous studies on the technological aspects more than the behavioral and psychological aspects (Lv et al., 2022). Therefore, it is necessary to stand on the elements of support and hindrance to Metaverse. These elements represent the compass of intellectual submission for individuals alike and the possibility of trying new things. This study aims to explore various aspects of metaverse. It investigates at both supportive and discouraging factors. To understand the possibility of adopting metaverse in the presence of these factors, the study focuses on social presence and social support as factors that support metaverse adoption. Social presence means a sense of actual presence in the digital space. It is an attraction mechanism used by metaverse technology in dealing with individuals (Sandberg et al., 2023; Hesselbarth et al., 2023). It enhances the possibility of actual interdependence and achieves the possibility of meaningful presence. At the same time, social support acts as an emotional and counseling informational guide. Social support is usually the various aspects of help and support provided by the social circle surrounding the individual. Therefore, this support is considered a driving force and enhances the possibility of entering the individual into the metaverse environment. While there is potential for supporting the adoption of metaverse, the metaverse environment is not without challenges. At the forefront of the challenges faced by the metaverse environment is users' concern about loss of privacy (Alnoor et al., 2023; Chew et al., 2023a). Loss of privacy means that data is public and not private property. Therefore, this is a great danger for individuals. It can lead to a decrease in the possibility of adopting metaverse. This study is a new addition in the world of digital transformation on the one hand, and business commerce on the other. Because it mixed the behavioral and technological elements to highlight what enhances or undermines the application of Metaverse in the business. The study is a guide for both policy makers and individuals, as it provides a statement to clarify a phenomenon that has been shrouded in obscurity for quite some time.

2 Virtual Applications

In order to clarify the process of influence and interaction between the variables of our study, it is important to present some of the main elements related to virtual applications. In other words, addressing these applications in terms of evidence and impact in practical reality. We can say that technology has played a major role in bringing everything together (Far et al., 2023). Rather, the matter has become more daring with the development of technology, which is living in virtual reality. Therefore, it represents a revolution in the field of interaction between individuals. These applications allow two realities to be included simultaneously, namely virtual reality (VR) and augmented reality (AR) (Jayawardena et al., 2023). There is a possibility to use these applications in various sciences, specializations, and even in different branches of one specialization. Virtual applications help individuals achieve immersive experiences. Therefore, these applications have ideal features and have significant impacts on various aspects of individuals' lives. One of its most important effects is the possibility of simulating reality via an electronic platform, glasses, or advanced screens (Durrani et al., 2022). This supports the possibility of entering into new experiences and thus creating spaces of communication. Communication through virtual applications crosses natural borders. Applications help create content for each user and then enable him to view, modify, and interact with his content. In addition to what was mentioned, there are also many benefits provided by virtual applications that we feel in our various daily experiences (Chew et al., 2023b).

3 Hypothesis Development

3.1 Social Presence and Adoption Metaverse Strategy

Social presence represents the state of existence that a person envisions about others (Khaw et al., 2022; AL-Fatlawey et al., 2021). This theory first appeared in psychology and communication studies (Cui et al., 2013). This theory deals with the perceived interactions between the interacting parties (Al-Abrrow et al., 2019). The various means of communication expanded the study of the theory in more depth. It has been shown recently that the abilities that characterize the tools of communication between the interacting parties can affect the perception of social presence (Bulu, 2012). It was applied in the field of e-marketing channels and social commerce using mobile applications (Attar et al., 2021; Alnoor et al., 2022). The social presence in the means of digital communication is a strategic tool to transform the hypothetical theory into a virtual reality (Mallmann and Maçada, 2021). It creates the feeling that you are interacting with real people. Depending on the simulation conducted by the digital signals between the interactions (AL-Abrrow, 2021). Therefore, it achieves the company's goal of adopting the metaverse strategy, by transferring users to a broad and shared digital space. Thus, the possibility of interacting with others as if they were actually present (Leong et al., 2020; Атия, 2016). Metaverse strategy is considered the most important dimension of the company in conveying a real interaction experience by adopting the most recent digital technologies. This requires a virtual environment based on digital tools that achieve the possibility of remarkable interactivity (Aburayya et al., 2023; Gatea and Marina, 2016). Therefore, social presence helps in understanding how to use digital tools to make interaction within the digital environment as tangible as possible (Heinze et al., 2016). For example, avatars bridge the gap between the real and virtual environment, facial expressions and gestures deepen the interaction process, and spatial sound technology creates a real interactive experience (Appel et al., 2020; Manhal et al., 2023). When users feel that there is a heavy presence in virtual reality, they are motivated to share their experiences and explore the experiences of others. The more dynamic the space, the greater the possibility of interaction (Onețiu, 2020). Thus, creating a constantly interactive community. This continuous interaction facilitates the possibility of feedback to improve attendance and interaction. Therefore, social presence is the key to adopting the metaverse strategy by enabling real social interaction resulting from a sense of connectedness in the digital environment (Voinea et al., 2022; Atiyah, 2023). It also instills conviction among users of the priority of digital interaction, as it is an embodiment of the truth, and even facilitates the possibility of obtaining broader and more detailed information. Hence the possibility of adopting the metaverse strategy in a vibrant and prosperous way (Hennig-Thurau et al., 2023). Thus, we assume:

Hypothesis 1. Social presence is positively connected with adoption metaverse strategy.

3.2 Social Support and Adoption Metaverse Strategy

Social support theory describes the process of positive interaction leading to support and feelings of appreciation. This theory was first studied in describing some social activities (Scharer, 2005). However, the multiple viewpoints in this theory allowed the possibility of studying the various processes of social influence, such as reinforcement, respect, appreciation, a sense of self, and engagement in constructive discussion societies (Hupcey, 1998; Atiyah, 2023). Social support is represented by various types of assistance provided by family members and friends, which contribute significantly to the formation of self-confidence and mental health, and thus greater flexibility in dealing with others (Aymen et al., 2019). With the development of digital commerce, social support has shifted towards a greater knowledge perspective, based on reality simulation applications (Khaw et al., 2022). Virtual reality is the most vital tool in coexistence with others from different places and the transmission of information about them (Izard et al., 2018). One of the most important of these applications is metaverse (Far and Rad, 2022). In this type of application, social support is crucial. As if the users live in the material world (Aria et al., 2023). Despite the nature of the interaction being purely digital, users need emotional and psychological support, assistance, and understanding within the digital environment (Ning et al., 2023; Atiyah, 2022). Therefore, when the organization formulates a metaverse strategy, it takes into account the potential for social support that can be provided to users (Koohang et al., 2023; Atiyah and Zaidan, 2022). Such as creating virtual communities for users with similar needs, allowing them to communicate and benefit from the required information. Generate virtual spaces for counseling and support, to facilitate access to assistance in a timely manner (Alfaisal et al., 2022). Expanding the discussion process by specialists, to achieve cooperation in obtaining an exceptional and real experience about the product (Buhalis et al., 2023). Social support processes help lay solid foundations for adopting the metaverse strategy, especially in the stages of encouraging users and enhancing their self-confidence, through emotional support and a sense of appreciation (Arpaci et al., 2022). Successful supportive discussions and answers to questions can create regularity for interaction in the metaverse environment. Therefore, social support facilitates everyone's involvement in the discussion in order to obtain the required information (Bian et al., 2021; Atiyah, 2020a, 2020b). The information provided in the virtual environment includes experiences from different places, so it helps in forming a future knowledge base for users about the product (Marabelli and Newell, 2022). From the foregoing, we come to the fact that social support is an important and influential element in adopting the Metaverse strategy. Through the volume of support, assistance, and solutions, as well as advice and enrichment that users provide to each other (Kozinets, 2022). Such broad and sympathetic processes give a highly flexible context that continuously nourishes the digital space and enriches its results. Therefore, we assume:

Hypothesis 2. Social support is positively connected with adoption metaverse strategy.

3.3 The Moderating Role of Privacy Risks

Privacy risks are the risks that the user of the electronic platforms may be exposed to in various forms. It consists of hacking his personal accounts and accessing his data (Ameen et al., 2022). Metaverse is one of the platforms with advanced digital technologies, using it may expose the user to identity theft and impersonation (Arpaci et al., 2022). Therefore, privacy risks raise the user's concern about unauthorized access to his intangible personal property (Rath and Kumar, 2021). Privacy risks play a pivotal role in the relationship between social presence and the adoption of the metaverse strategy (Chakraborty et al., 2023). The higher the privacy risk, the less users feel social presence in the metaverse environment, and vice versa. If the user feels that the company is making a great effort to protect his privacy within the metaverse space, the user will be convinced that his data is protected and that there are no problems with its use (Nadon et al., 2018; Atiyah, 2020a, 2020b). Thus, users are more present. This leads to the success of adopting the metaverse strategy. The company must employ its advertising capabilities in marketing the metaverse strategy as a safe application with the ability to protect user privacy (Gadekallu et al., 2022). Without that, the user's presence in the metaverse space will depend on personal conviction and the allegations made by competing companies. If the user realizes the weakness of privacy protection, unlikely to enter the metaverse environment, and therefore does not have a presence and is at least apprehensive and non-interactive, which leads to his insecurity, as the potential risks for him are more than the benefits (Hazan et al., 2022; Njoku et al., 2023). Hence, we suppose:

Hypothesis 3. The privacy risks moderate the relationship between social presence and adopt a metaverse strategy.

Privacy risks are a threat to users of modern digital technologies (Ameen et al., 2022). This threat is to break into and obtain user data. In addition to the possibility of displaying the privacy of others and transferring them from one place to another (Quach et al., 2022). Access to data is usually through illegal means such as fraud and hacking. Highly sensitive data such as personal data is an important element in expressing user privacy (Makkonen et al., 2019). The metaverse application is one of the applications that causes great concern among potential users. Privacy risks play a major role in hindering its implementation (Arpaci et al., 2022). If the level of privacy risk is low and there is social support, then the metaverse strategy can be adopted successfully. When individuals feel that companies care about the privacy of their data, they have the desire to deal with metaverse and try it (Anshari et al., 2022). Users start interacting with others and get social support through the social networks provided by metaverse. Social support enhances self-confidence and allows users to experience new activities in the metaverse environment (Benrimoh et al., 2022). On the other hand, if the privacy risks are high, there is a great concern among the potential users in entering the metaverse space, and therefore companies cannot adopt the metaverse strategy. Furthermore, individuals realize that companies do not pay enough attention to the privacy of their data, worry about using the metaverse and treat social support with extreme caution (Bibri and Allam, 2023). Consequently, their level of confidence is very low in the information that social networking sites provide them (Dwivedi et al., 2021). It can be said that the moderated role of privacy risks in the relationship between social support and the adoption of the metaverse strategy is represented by nature of dealing with user privacy. It is true that the user needs social support, but this is not enough to enter into a complex digital environment. Therefore, a high level of reassurance must be provided to the user so that has sufficient conviction that the user is interacting in a safe environment that is keen on his privacy (Letafati and Otoum, 2023). So, we assume:

Hypothesis 4. The privacy risks moderate the relationship between social support and adopt a metaverse strategy.

As shown in Fig. 1, the conceptual framework of this study discovers a relationship between variables. Effect of social presence and social support (interpretive variables) on adoption of metaverse strategy (responsive variable). And moderative effect of privacy risks (moderative variable). This framework was constructed based on (Sekaran & Bougie, 2016: 75).

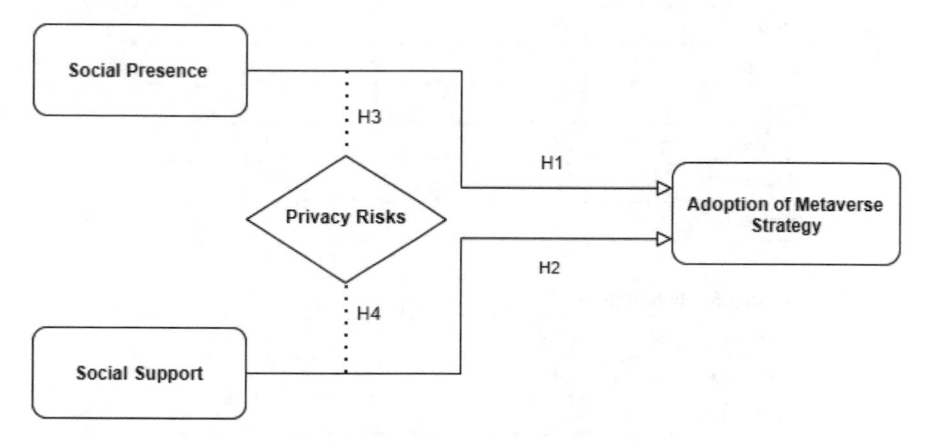

Fig. 1. Conceptual Framework.

4 Research Methodology

This study used a survey method. The data of the study was collected directly from respondents by questionnaire technique. The study's sample was some of the users of electronic platforms (Instagram, Facebook, LinkedIn) that were randomly selected. As google form was formulated for this purpose. To meet the requirements of the scale respondents must have the following characteristics: 1) their age must be more than (30) years, to obtain clear and accurate answers, 2) they must have used social commerce platforms for three years or more, to have sufficient experience. The study adopted English survey items, therefore translated into Arabic according to Brislin's procedure (1970). To ensure the validity of the scale a back-translation tool was used by a specialist in both English and Arabic. Furthermore, the translated items have been reviewed by managers and academics to achieve their applicability and comprehension. The questionnaire was sent to 250 responders. Only 223 of the responders completed and resend it. 13 of the questionnaires were ineligible. Therefore, 210 were adopted as the study's data. The demographic characteristics of the participants are shown in Table 1. 30–39 years represent the highest percentage .30 of the age of the study sample. While ≥60 represents the lowest percentage .18. Undergraduate was held by .85% of respondents, and .15% received a post-graduate degree. 7–10 years represent the highest percentage .68 of the experience of the study sample. While 3–6 represents the lowest percentage .9. The typical respondent lasted for roughly 9 years. Most participants were between the ages of 30 and 39.

Table 1. Demographics of the participations.

Demographics	Frequency (n = 210)	Percent %
Gender		
Male	*150*	*.71*
Female	*60*	*.29*
Age		
30–39	*63*	*.30*
40–49	*55*	*.26*
50–59	*54*	*.26*
≥60	*38*	*.18*
Academic qualifications		
Undergraduate	*178*	*.85*
Post-graduate	*32*	*.15*
Experience		
3–6	*18*	*.09*
7–10	*144*	*.68*
≥11	*48*	*.23*

The four original variables in this study involve social presence and social support as independent variables, privacy risks as a moderative variable, and adoption metaverse strategy as the dependent variable. To measure these variables this study used well-established scales from available literature. To measure social presence, we used 12 items and to measure social support we used 6 items, based on Khaw et al. (2022). As well as, used 3 items for the adoption metaverse strategy, adopted from (Arpaci and Bahari, 2023). In contrast, privacy risks were measured with 4 items adapted from (Ameen et al., 2022). A 7-point scale (1 = strongly disagree, 7 = strongly agree) was used to operationalize all the concepts.

5 Data Analysis

According to Hair et al. (2010), the objective of the measurement model is to evaluate the validity and reliability of the generated measures. To seek convergent validity must evaluate the average extracted variance and composite reliability. As shown in Table 2 some estimates of these items were deleted because less than 0.50 and others were more than 0.50, as mentioned by hair et al. (2010), and the average variance extracted (AVE) values of all latent variables were from 0.696 to 0.743, which was normal. Additionally, the outer loadings of all latent variables ranged between 0.778 and 0.919, more than the acceptable value of 0.70 as referred by Hair et al. (2017). The latent variables' composite reliabilities also ranged between 0. 840 and 0. 888. These results show that the measurement scales utilized in the model are great reliability (Hair et al., 2010).

Table 2. Result of measurement model.

Variables	Items	Loading factors	CR	AVE
Social presence Khaw et al. (2022)	Human warmth in the web of the s-commerce seller	0.87	0.873	0.696
	Human sensitivity in the web of the s-commerce seller social presence of Communication	0.814		
	Character of sellers by interacting with them via s-commerce	0.818		
Social support. Khaw et al. (2022)	Some people on the s-commerce site are on my side with me if any issue	0.827	0.888	0.725
	Some people on the s-commerce site comforted and encouraged me if any issue	0.869		
	Some people on the s-commerce site expressed interest and concern in my well-being if any issue Mobile-wallet resistance	0.859		
Metaverse strategy.Arpaci and Bahari, 2022	Will always try to use Metaverse for main purposes	0.821	0.852	0.743
	I tend to use Metaverse for secondary purposes frequently	0.901		

(continued)

Table 2. (*continued*)

Variables	Items	Loading factors	CR	AVE
Privacy risks. Ameen et al., 2022	I am sensitive about giving out information regarding my preferences	0.919	0.84	0.725
	I am concerned about anonymous information that is collected about me via different technologies while shopping in the mall	0.778		

To obtain results of discriminant validity the present study used Fornell and Larcker's (1981) method. As shown in Table 3 the square root of the AVE of each factor is more than the correlation estimates of the factors. This illustrates all the factors are clearly different from one another, meaning that each factor is rare and explains phenomena not represented by other factors in the model (Hair et al., 2010).

Table 3. Discriminant validity.

Variables	MS	PR	SS	SP
Metaverse strategy	0.862			
Privacy risks	0.737	0.851		
Social presence	0.418	0.307	0.834	
Social support	0.630	0.530	0.349	0.852

In this study, there are four hypotheses were hypothesized to achieve the study objective. By using a bootstrapping method based on PLS-SEM (Hair et al., 2017). We tested hypotheses, parameter estimates for statistical significance, and coefficient values. The bootstrapping way with 5000 bootstrap re-sampling and bias-corrected confidence periods was used to test the significance of the path coefficients. Table 4. Shows the results of the structural model assessment.

The findings in Table 4 report a direct effect, but there was high evidence that all statistical hypotheses on direct influence were valid. Regarding the indirect effect assumptions, the findings explained that the moderator factor plays a fully moderating role in the interaction between the dependent and the independent factors (Fig. 2).

Table 4. Hypotheses Test.

Direct path	O	M	SD	O	P
Social presence → Metaverse strategy	0.152	0.154	0.038	3.989	0.000
Social support → Metaverse strategy	0.295	0.296	0.047	6.230	0.000
Privacy risks × Social presence → Metaverse	0.411	0.012	0.032	12.844	0.000
Privacy risks × Social support → Metaverse	0.213	0.013	0.035	6.086	0.016

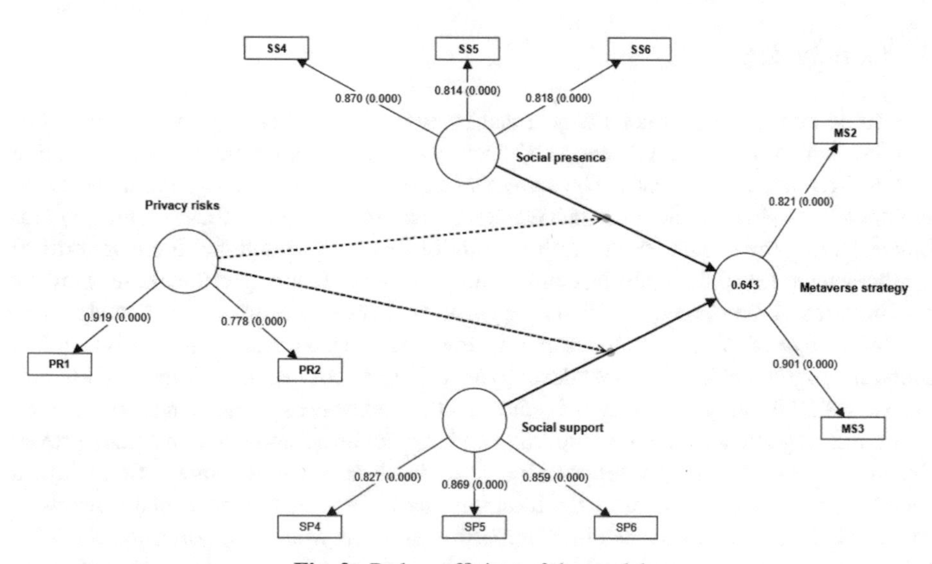

Fig. 2. Path coefficient of the model.

6 Discussion

The results shown in Table 4 allow us to understand the nature of the influence of the study model variables. Regarding social presence, the results of the study showed that the social presence that the individual perceives of others around him is an important element in activation with digital space (H2). This result is consistent with (Mallmann and Maçada, 2021) statement. Which states that the social presence in the means of digital communication is a strategic tool to transform the hypothetical theory into a virtual reality. Also, there is an effect of social support in adopting the Metaverse strategy, this result supports H1. And agrees with what (Khaw et al., 2022) indicated about the importance of social support in entering the virtual world and interacting with it. Moreover, shows the reality of social support through various social networks. Hence the important role that social support plays in creating a spirit of interaction among individuals in the digital space. The results of the study support the hypothesis that states the interactive role of privacy risks in the relationship between social presence and metaverse adoption, (H3). This greatly reduces the possibility of adopting the Metaverse strategy. Although social presence is effective element in the Metaverse environment, but the risks of privacy represent an obstacle in this presence, as agrees with opinions of Hazan et al. (2022) and Njoku et al. (2023). Finally, the results show clear support for the H4. Meaning that privacy risks reduce the likelihood of adopting the Metaverse strategy. These results explain the emergence of privacy risks as a phenomenon that reduces the possibility of expanding the metaverse environment in businesses. That is what was indicated by (Bibri and Allam, 2023). Therefore, it is important to deal with these results seriously and think about them in-depth. This will be explained in the next part of this study.

7 Conclusion

The study aimed to construct a conceptual framework based on two main elements of support, which are social presence and social support. Another element of frustration or obstruction is privacy risks. The study was directed towards measuring the extent to which it is possible to adopt the metaverse strategy in the light of these elements combined. During the excursion into the literature of this study found that there are critical opinions regarding the supporting elements of metaverse. This means that these elements are important to be studied in various organizational environments. Unfortunately, there was no study that dealt with metaverse in Iraq. Although this strategy can be applied like other digital applications that develop the ability of companies to interact with their customers. This study has come to the fact that the metaverse strategy cannot be adopted without taking into account the supporting and discouraging elements. The study proved that social presence strongly supports the individual's ability to feel interaction within a secure environment. In the sense the information obtained by the individual is identical to reality. Social support is the other supporting element. Social support plays a pivotal role in obtaining information and facilitating the individual's access to the digital world of metaverse. But the privacy risk is the frustrating element of access to the metaverse environment. Thus, it hinders its adoption as a strategy by companies. Because the individual remains in a spiral of anxiety and doubts about his personal data and privacy. These results lead to the importance of taking into consideration these elements. Either supportive or obstructive. Policy makers should take seriously the importance of thinking about resolving this somewhat complex phenomenon.

References

Aburayya, A., et al.: SEM-machine learning-based model for perusing the adoption of metaverse in higher education in UAE. Int. J. Data Netw. Sci. **7**(2), 667–676 (2023)

AL-Abrrow, H., et al.: Understanding employees' responses to the COVID-19 pandemic: the attractiveness of healthcare jobs. Glob., Bus. Organ. Excellence **40**(2), 19–33 (2021)

Al-Abrrow, H., Alnoor, A., Abbas, S.: The effect of organizational resilience and CEO's narcissism on project success: Organizational risk as mediating variable. Organ. Manag. J. **16**(1), 1–13 (2019)

Alfaisal, R.M., Zare, A., Alfaisal, A.M., Aljanada, R., Abukhalil, G.W.: The acceptance of metaverse system: a hybrid SEM-ML approach. Int. J. Adv. Appl. Comput. Intell. **1**(1), 34–44 (2022)

AL-Fatlawey, M.H., Brias, A.K., Atiyah, A.G.: The role of strategic behavior in achievement the organizational excellence. Analytical research of the manager's views of Ur State Company at Thi-Qar Governorate. J. Admin. Econ. **10**(37). (2021)

Alnoor, A., Khaw, K.W., Chew, X., Abbas, S., Khattak, Z.Z.: The influence of the barriers of hybrid strategy on strategic competitive priorities: evidence from oil companies. Glob. J. Flex. Syst. Manag. **24**(2), 179–198 (2023)

Alnoor, A., et al.: How positive and negative electronic word of mouth (eWOM) affects customers' intention to use social commerce? A dual-stage multi group-SEM and ANN analysis. Int. J. Hum.–Comput. Interact. 1–30 (2022)

Ameen, N., Hosany, S., Paul, J.: The personalisation-privacy paradox: consumer interaction with smart technologies and shopping mall loyalty. Comput. Hum. Behav. **126**, 106976 (2022)

Anshari, M., Syafrudin, M., Fitriyani, N.L., Razzaq, A.: Ethical responsibility and sustainability (ERS) development in a metaverse business model. Sustainability **14**(23), 15805 (2022)

Appel, G., Grewal, L., Hadi, R., Stephen, A.T.: The future of social media in marketing. J. Acad. Mark. Sci. **48**(1), 79–95 (2020)

Aria, R., Archer, N., Khanlari, M., Shah, B.: Influential factors in the design and development of a sustainable Web3/Metaverse and its applications. Future Internet **15**(4), 131 (2023)

Arpaci, I., Karatas, K., Kusci, I., Al-Emran, M.: Understanding the social sustainability of the Metaverse by integrating UTAUT2 and big five personality traits: a hybrid SEM-ANN approach. Technol. Soc. **71**, 102120 (2022)

Atiyah, A.G.: Impact of knowledge workers characteristics in promoting organizational creativity: an applied study in a sample of Smart organizations. PalArch's J. Archaeol. Egypt/Egyptol. **17**(6), 16626–16637 (2020)

Atiyah, A.G.: The effect of the dimensions of strategic change on organizational performance level. PalArch's J. Archaeol. Egypt/Egyptol. **17**(8), 1269–1282 (2020)

Atiyah, A.G.: Effect of Temporal and spatial myopia on managerial performance. J. La Bisecoman **3**(4), 140–150 (2022)

Atiyah, A.G.: Power Distance and Strategic Decision Implementation: Exploring the Moderative Influence of Organizational Context

Atiyah, A.G.: Strategic Network and Psychological Contract Breach: The Mediating Effect of Role Ambiguity.

Atiyah, A.G., Zaidan, R.A.: Barriers to using social commerce. In: Alnoor, A., Wah, K.K., Hassan, A. (eds.) Artificial Neural Networks and Structural Equation Modeling, pp. 115–130. Springer, Singapore (2022). https://doi.org/10.1007/978-981-19-6509-8_7

Attar, R.W., Shanmugam, M., Hajli, N.: Investigating the antecedents of e-commerce satisfaction in social commerce context. Brit. Food J. **123**(3), 849–868 (2021)

Aymen, R.A., Alhamzah, A., Bilal, E.: A multi-level study of influence knowledge management small and medium enterprises. Pol. J. Manag. Stud. **19**(1), 21–31 (2019)

Benrimoh, D., Chheda, F.D., Margolese, H.C.: The best predictor of the future—the metaverse, mental health, and lessons learned from current technologies. JMIR Mental Health **9**(10), e40410 (2022)

Bian, Y., Leng, J., Zhao, J.L.: Demystifying metaverse as a new paradigm of enterprise digitization. In: Wei, J., Zhang, L.J. (eds.) BigData 2021. LNCS, vol. 12988, pp. 109–119. Springer, Cham (2021). https://doi.org/10.1007/978-3-030-96282-1_8

Bibri, S.E., Allam, Z.: The Metaverse as a virtual form of data-driven smart cities: the ethics of the hyper-connectivity, datafication, algorithmization, and platformization of urban society. Comput. Urban Sci. **2**(1), 22 (2023)

Buhalis, D., Leung, D., Lin, M.: Metaverse as a disruptive technology revolutionising tourism management and marketing. Tour. Manag. **97**, 104724 (2023)

Bulu, S.T.: Place presence, social presence, co-presence, and satisfaction in virtual worlds. Comput. Educ. **58**(1), 154–161 (2012)

Chakraborty, D., Patre, S., Tiwari, D.: Metaverse mingle: discovering dating intentions in metaverse. J. Retail. Consum. Serv. **75**, 103509 (2023)

Chan, S.H.M., Qiu, L., Xie, T.: Understanding experiences in metaverse: how virtual nature impacts affect, pro-environmental attitudes, and intention to engage with physical nature. Comput. Hum. Behav. 107926. (2023)

Chew, X., Alharbi, R., Khaw, K.W., Alnoor, A.: How information technology influences organizational communication: the mediating role of organizational structure. PSU Res. Rev. (2023a)

Chew, X., Khaw, K.W., Alnoor, A., Ferasso, M., Al Halbusi, H., Muhsen, Y.R.: Circular economy of medical waste: novel intelligent medical waste management framework based on extension linear Diophantine fuzzy FDOSM and neural network approach. Environ. Sci. Pollut. Res. 1–27 (2023b)

Cui, G., Lockee, B., Meng, C.: Building modern online social presence: a review of social presence theory and its instructional design implications for future trends. Educ. Inf. Technol. **18**, 661–685 (2013)

Durrani, S., et al.: The virtual vision of neurosurgery: how augmented reality and virtual reality are transforming the neurosurgical operating room. World Neurosurgery (2022)

Dwivedi, Y.K., et al.: Metaverse beyond the hype: multidisciplinary perspectives on emerging challenges, opportunities, and agenda for research, practice and policy. Int. J. Inf. Manag. **66**, 102542 (2021)

Far, S.B., Rad, A.I.: Applying digital twins in metaverse: user interface, security and privacy challenges. J. Metaverse **2**(1), 8–15 (2022)

Far, S.B., Rad, A.I., Bamakan, S.M.H., Asaar, M.R.: Toward Metaverse of everything: opportunities, challenges, and future directions of the next generation of visual/virtual communications. J. Netw. Comput. Appl. 103675 (2023)

Gadekallu, T.R., et al.: Blockchain for the metaverse: a review. arXiv preprint arXiv:2203.09738 (2022)

Gatea, A.A., Marina, V.: Higher education funding in Iraq in terms of the experience of particular developed countries. Int. J. Adv. Stud. **6**(1), 8–17 (2016)

Hazan, E., Kelly, G., Khan, H., Spillecke, D., Yee, L.: Marketing in the metaverse: an opportunity for innovation and experimentation. McKinsey Q. (2022)

Heinze, A., Fletcher, G., Cruz, A.: Digital and Social Media Marketing: A Results-Driven Approach. Routledge, London (2016)

Hennig-Thurau, T., Aliman, D.N., Herting, A.M., Cziehso, G.P., Linder, M., Kübler, R.V.: Social interactions in the metaverse: framework, initial evidence, and research roadmap. J. Acad. Mark. Sci. **51**(4), 889–913 (2023)

Hesselbarth, I., Alnoor, A., Tiberius, V.: Behavioral strategy: a systematic literature review and research framework. Manag. Decis. (2023)

Hupcey, J.E.: Clarifying the social support theory-research linkage. J. Adv. Nurs. **27**(6), 1231–1241 (1998)

Izard, S.G., Juanes, J.A., García Peñalvo, F.J., Estella, J.M.G., Ledesma, M.J.S., Ruisoto, P.: Virtual reality as an educational and training tool for medicine. J. Med. Syst. **42**, 1–5 (2018)

Jayawardena, N.S., Thaichon, P., Quach, S., Razzaq, A., Behl, A.: The persuasion effects of virtual reality (VR) and augmented reality (AR) video advertisements: a conceptual review. J. Bus. Res. **160**, 113739 (2023)
Khaw, K.W., et al.: Modelling and evaluating trust in mobile commerce: a hybrid three stage Fuzzy Delphi, structural equation modeling, and neural network approach. Int. J. Hum.-Comput. Interact. **38**(16), 1529–1545 (2022)
Koohang, A., et al.: Shaping the metaverse into reality: a holistic multidisciplinary understanding of opportunities, challenges, and avenues for future investigation. J. Comput. Inf. Syst. **63**(3), 735–765 (2023)
Kozinets, R.V.: Immersive netnography: a novel method for service experience research in virtual reality, augmented reality and metaverse contexts. J. Serv. Manag. **34**(1), 100–125 (2022)
Leong, L.Y., Hew, T.S., Ooi, K.B., Chong, A.Y.L.: Predicting the antecedents of trust in social commerce–a hybrid structural equation modeling with neural network approach. J. Bus. Res. **110**, 24–40 (2020)
Letafati, M., Otoum, S.: On the privacy and security for e-health services in the metaverse: an overview. Ad Hoc Netw. 103262 (2023)
Lv, Z., Xie, S., Li, Y., Hossain, M.S., El Saddik, A.: Building the metaverse by digital twins at all scales, state, relation. Virtual Real. Intell. Hardw. **4**(6), 459–470 (2022)
Makkonen, P., Lampropoulos, G., Siakas, K.: Security and privacy issues and concerns about the use of social networking services. In: E-Learn: World Conference on E-Learning in Corporate, Government, Healthcare, and Higher Education, pp. 457–466. Association for the Advancement of Computing in Education (AACE) (2019)
Mallmann, G.L., Maçada, A.C.G.: The mediating role of social presence in the relationship between shadow IT usage and individual performance: a social presence theory perspective. Behav. Inf. Technol. **40**(4), 427–441 (2021)
Manhal, M., Al-khalidi, A., Hamad, Z.: Strategic network: managerial myopia point of view. Manag. Sci. Lett. **13**(3), 211–218 (2023)
Marabelli, M., Newell, S.: Everything you always wanted to know about the metaverse*(* But were afraid to ask). In: Academy of Management Annual Meeting (2022)
Mourtzis, D., Panopoulos, N., Angelopoulos, J., Wang, B., Wang, L.: Human centric platforms for personalized value creation in metaverse. J. Manuf. Syst. **65**, 653–659 (2022)
Nadon, G., Feilberg, M., Johansen, M., Shklovski, I.: In the user we trust: unrealistic expectations of Facebook's privacy mechanisms. In: Proceedings of the 9th International Conference on Social Media and Society, pp. 138–149 (2018)
Ning, H., et al.: A survey on the metaverse: the state-of-the-art, technologies, applications, and challenges. IEEE Internet Things J. (2023)
Njoku, J.N., Nwakanma, C.I., Amaizu, G.C., Kim, D.S.: Prospects and challenges of Metaverse application in data-driven intelligent transportation systems. IET Intel. Transp. Syst. **17**(1), 1–21 (2023)
Onețiu, D.D.: The impact of social media adoption by companies. Digital transformation. Studia Universitatis Vasile Goldiş, Arad-Seria Ştiinţe Economice **30**(2), 83–96 (2020)
Pellegrino, A., Stasi, A., Wang, R.: Exploring the intersection of sustainable consumption and the Metaverse: a review of current literature and future research directions. Heliyon (2023)
Quach, S., Thaichon, P., Martin, K.D., Weaven, S., Palmatier, R.W.: Digital technologies: tensions in privacy and data. J. Acad. Mark. Sci. **50**(6), 1299–1323 (2022)
Rath, D.K., Kumar, A.: Information privacy concern at individual, group, organization and societal level-a literature review. Vilakshan-XIMB J. Manag. **18**(2), 171–186 (2021)
Sandberg, H., Alnoor, A., Tiberius, V.: Environmental, social, and governance ratings and financial performance: evidence from the European food industry. Bus. Strateg. Environ. **32**(4), 2471–2489 (2023)

Scharer, K.: Internet social support for parents: the state of science. J. Child Adolesc. Psychiatr. Nurs. **18**(1), 26–35 (2005)
Sekaran, U., Bougie, R.: Research Methods For Business: A Skill-Building Approach. Wiley, Hoboken (2016)
Voinea, G.D., et al.: Study of social presence while interacting in metaverse with an augmented avatar during autonomous driving. Appl. Sci. **12**(22), 11804 (2022)

Employing Metaverse Technologies to Improve the Quality of the Educational Process

Abdulridha Nasser Mohsin, Munaf abdulkadim Mohammed, and Marwa Al-Maatoq(✉)

Technical College of Management, Sothern Technical University, Basra, Iraq
{dr.abdnaser,manaf.alqattan,marwa.mousa993}@stu.edu.iq

Abstract. The term Metaverse denotes the forthcoming iteration of the internet, which is a recently introduced concept within our scholarly community, particularly in the realm of higher education. The objective of this study is to investigate the influence of Metaverse technologies on improving the caliber of higher education. This text further explores the analytical presentation of previous studies on Metaverse technologies, including virtual reality VR, augmented reality AR, and mixed reality MR. Besides, it examines the quality of the educational process, focusing on the roles of teachers, students, and the curriculum. The analysis is conducted using a theoretical review methodology. This study goals to scan the practical applications of Metaverse technology in enhancing the learning process by analyzing real-life case studies. As a result, the paper arrived at several conclusions and recommendations, with the most significant being that Metaverse technologies played a crucial part in enhancing the educational process by introducing innovative e-learning technologies.

Keywords: Metaverse · educational process · including virtual reality · mixed reality · augmented reality

1 Introduction

The phrase Metaverse is a fusion or amalgamation of "Meta", signifying "beyond" or "after", and "Verse", referring to "the universe" or "the world". Therefore, Metaverse denotes a realm that surpasses reality. Alternatively, it refers to a cosmos consisting of an interconnected system of virtual realms that enable users to engage in a deeply immersive online encounter (Recker et al., 2021). Moreover, with the advancement of virtual reality systems and their integration with networked reality technologies, there is immense potential for these technologies to bring about revolutionary changes in various domains such as education, distant work, marketing, economics, and the entertainment industry. Furthermore, these technologies have given rise to a novel and sophisticated mode of information exchange known as the Metaverse (Kaddoura, and Al Husseiny, 2023; Al-Abrrow et al., 2021). From a scholastic view, both the business and commercial sectors require highly educated individual who can effectively address the complex challenges of Metaverse environments. Consequently, this necessitates the adoption of

M. Al-Emran et al. (Eds.): IMDC-IST 2024, LNNS 876, pp. 159–174, 2023.
https://doi.org/10.1007/978-3-031-51300-8_11

new models of organizational and leadership management. Furthermore, it is necessary to provide a description of human activity in Metaverse environments. It is essential to conduct investigations inside an educational setting to ascertain the disparities between behavior observed in this controlled context and that exhibited in the real world. Likewise, education holds great potential as one of the most attractive uses of Metaverse in the foreseeable future. Educational institutions can derive advantages from adaptable platforms that facilitate seamless interaction among professors, students, and people, transcending the limitations of physical classrooms. Consequently, the Metaverse effectively encompasses authentic learning settings (Al-Adwan et al., 2023; Albahri et al., 2021). Hence, the issue at hand in this study can be succinctly encapsulated in the ensuing query: What is the effect of employing Metaverse technologies on enhancing the quality process of the educational system?

The paper aims to accomplish a specific set of goals. Firstly, it is essential to diagnose the nature of the influence relationship and the correlation between Metaverse technologies and the educational process quality. In addition, it is crucial to unveil the true nature of educational facilities and acquire the knowledge of effectively utilizing Metaverse technology to enhance the educational process. Ultimately, the objective is to discover methods for enhancing the educational process through the utilization of Metaverse technologies. This work holds significance as it makes a important contribution to the current knowledge on the Metaverse. It provides valuable insights that can assist educational institutions in leveraging Metaverse technology to improve the overall quality of the educational process. In addition, this study concentrates on essential and pertinent problems that have been neglected by prior research. Specifically, it examines how to connect instructional infrastructures with modern technological breakthroughs and improve their overall efficiency. Furthermore, this present study offers the potential for future studies and other fields to derive benefits. It also provides senior management with valuable insights into the significance of Metaverse technologies and strategies for achieving exceptional performance. In order to achieve the intended objective, the current study is structured in the following manner: Sect. 2 provides a comprehensive literature analysis encompassing the primary definitions of the essential concepts. The Methodology and depiction theoretical review analysis are outlined in Sect. 3. Section 4 discusses the conclusions and future recommendations of this study.

2 Literature Review

2.1 The Concept of Metaverse Technologies in the Educational Process

The "Metaverse" technology was presented by Stephenson in 1992 in his science fiction book. Metaverse is a concept that facilitates interactions and daily communication among people. It can be described as a realm that improves physical space and reality, integrating the electronic and physical universes. Individuals can imagine digital representations of different aspects of the physical world and virtual elements that do not exist in reality for different purposes (Alfaisal et al., 2022; AL-Fatlawey et al., 2021). Following Zuckerberg's open announcement of the Metaverse program late months in 2021, the term "Metaverse" gained significant attention. Educators and academics started discussing various plans and implementing strategies for incorporating the Metaverse into their educational procedures. The increasing fascination with the educational environment arises from several opportunities, such as virtual space, which provides realistic simulations that have the ability to improve the social dimension of education and instruction (Tlili et al., 2022; Alnoor et al., 2022). Online education has gained interest as technology has developed pervasive in human lives, with the Metaverse being one of the innovations that emerged in the late nineties and continues to be updated and improved to adapt to current changes. However, many instructors and learners do not fully embrace it. They have the expertise and abilities to employ the Metaverse effectively in revolutionizing the educational process. Previous research findings indicate that students derive great satisfaction from using Metaverse as a means of instruction, enabling them to grasp various subjects more comprehensively compared to conventional curriculum-based and in-person education. The findings suggest that Metaverse holds significant promise for future exploration in the educational sector, as it facilitates the improvement of technological skills and greatly enhances student engagement. However, to mitigate any potential drawbacks, students require instruction from both teachers and parents (Onggirawan et al., 2023).

2.2 Characteristics of Metaverse Technologies in the Educational Process

The term "Metaverse" is a relatively new concept that encompasses various aspects of individual development, including online interaction technologies, electronic channels, and decentralization, the integration of reality and virtually, and human-computer interaction. In the context of education, these features have been highlighted by researchers such as (Lin et al., 2022); (Contreras et al., 2022; Alsalem et al., 2022).

a. Decentralization, inherent in Block chain technology, disrupts the conventional social process by shifting the creation of the Metaverse from a particular group to all individuals, fostering engagement and equal invention.
b. The term "Internet of Things" (IoT) refers to the widespread adoption of Internet connectivity for everyday objects. It entails linking the real and virtual worlds so that people can move freely between them regardless of time or place. In the Metaverse, individuals contemplate their identity (including their social connections in both worlds) and ponder the potential impact of this virtual reality on their manner of life.
c. The Metaverse's fundamental purpose is human-computer communication, wherein the level of efficiency directly dictates the boundaries of personal significance and the utility of the Metaverse for users.
d. Digital reality encompasses AR, VR, MR, and 3-D video, providing consumers with a comprehensive experience. Such innovations enable users to overcome the limitations of period and space, as immersive technology can intensify comprehension and improve the learning experience for students.
e. The monetary value of Metaverse in education surpasses that of conventional education due to the capability of people to create diverse digital services and the increased flexibility and convenience of actions between people, such as transitioning from the physical realm to the digital realm and vice versa.

Education, and more narrowly, the application of the Metaverse in learning, has been the primary focus of research. A significant amount of literature has been dedicated to exploring the use of VR in learning, given its dynamic nature and its potential to enhance educational processes. The Metaverse offers numerous benefits in the educational process because of its updated and organized materials, as well as its seamless communication among teachers and students, as follows (Onecha et al., 2023; Atiyah, 2020); (Hwang and Chien, 2022; Fadhil et al., 2021):

a. The Metaverse has the capability to elucidate intricate or intangible ideas and to generate comprehensive and comprehensive educational experiences by means of active and engaging action and interactions.
b. The Metaverse provides opportunities for remote social education and knowledge interchange, aligning with the essential features of international education.
c. The integration of the Metaverse in learning allows for unlimited temporal and spatial boundaries, customization, and the avoidance of academic dishonesty through the usage of Block chain technology.
d. The Metaverse is transforming the field of learning through utilizing sophisticated information analysis techniques and teaching strategies, educators can offer intricate and tailored assignments to pupils, fostering their autonomy and self-reliance.
e. The Metaverse platform offers a comprehensive and unified educational experience that fosters responsible usage values among students.

With the advent of Metaverse, a completely novel period of technological advancements in education has begun. In order to incorporate technology effectively into their teaching, educators must thoroughly understand its benefits and drawbacks, and take necessary measures to enhance its effectiveness while minimizing potential risks. The challenges of Metaverse are outlined as follows (Fitria and Simbolon, 2022; Gatea and Marina, 2016):

a. Metaverse necessitates significant technological usage due to the production of intricate graphics and high-quality visuals.
b. The acquisition of the technologies that facilitate the Metaverse incurs significant expenses.
c. The potential for social and cultural transformation resulting from the Metaverse is boundless.
d. The Metaverse is comparatively less significant than the virtual environment and lacks the same level of connectivity and activity.

2.3 Indicators of Metaverse Technologies in the Educational Process

For academic institutions in order to virtualize and effectively utilize the Metaverse technology, and enhance the educational process they must consider the following indicators (Hassanzadeh, 2022; Hamid et al., 2021):

a. Specialized infrastructure is necessary to engage in the world of Metaverse.
b. Establish community support and dedication from the Metaverse generation by implementing effective virtual community awareness strategies.

c. Modifying the educational institution's management framework to align with this emerging Metaverse, as conventional structures may encounter significant limitations in their ability to engage effectively.
d. Utilize the powers of educational autonomy to regulate the Metaverse and safeguard public interests during the promotion of the Metaverse.
e. Enhancing education across all grade levels and subject areas by implementing a curriculum that instructs the Meta generation and equips them for a proficient and engaged role in the Metaverse.

2.4 Types of Metaverse Technologies in Educational Process

Metaverse is an innovative internet service that integrates VR and, two advanced innovations. This amalgamation is anticipated to exert a substantial impact on the realm of higher education, as it is still emerging. The implementation of Virtual and Augmented Reality (VAR) technology as an institutional platform in university education is still in its initial phase. This technology enables realistic learning experiences in locations that learners are unable to realistically access. It achieves this by the use of 3D models and interactive 360° movies. The outcomes of implementing this innovative technology Academic organizations are assured that investing in the necessary infrastructure for virtual and augmented reality technology, as well as mixed reality technology that incorporates both, is a wise investment for the subsequent years (Marks & Thomas, 2022; Khaw et al., 2022).

The inception of VR may be traced back to the 1960s witch is a device that created a realistic experience by employing a moving 3D image to excite the senses of sight, hearing, smell, and touch. VR has primarily been used in the entertainment industry, but since the 1980s, research has been carried out on its programs and effectiveness in educational institutions and training. The integration of VR technology and education offers a new approach that enhances conventional methods. This approach enhances learners' interest in learning by providing visual experiences and communicating information (Smutny, 2022). The technological aspects of VR might be categorized into two types: low immersion and high immersion. Low immersion VR, also known as desktop VR, Entails the presentation of a simulated environment on a computer screen and engaging with it through the utilization of a mouse, keyboard, or joystick. In this particular form of virtual reality, individuals possess a comprehensive comprehension of the actual physical world, hence creating a secure atmosphere with few hazards. On the other hand, high-immersion VR requires more advanced equipment and interaction. It typically involves using a head-mounted exhibit with excellent quality graphics and isolated lenses for both eyes, along with audio delivered through headphones. This type of VR allows the user to fully participate to the simulated environment. It is worth noting that high-immersion VR is more interactive and requires expensive peripherals, while low-immersion VR is less immersive and does not include as much communication (Liberman & Dubovi, 2023; Manhal et al., 2023).

AR is a technology that uses mobile devices like smartphones and tablets to superimpose digital content onto a user's view of the physical world. Furthermore, this innovative technology has the potential to be utilized in the realm of learning, as it empowers individuals to enhance the physical environment with digital content. Additionally, it enriches students' real-world experiences by seamlessly connecting digital resources with the physical environment. This is achieved by matching information from the user's virtual environment with physical objects in the immediate area. Conversely, augmented reality encompasses a substantial amount of virtual data that enhances the physical environment and is regularly accessible (Elbyaly & Elfeky, 2023). In addition, the researcher aims to highlight the findings of the study conducted by (Hidajat, 2023), which are as follows:

1. Augmented reality (AR) enhances pupils' cognitive performance by addressing their self-ability and social skills challenges.
2. The key components in the advancement of (AR) apps are the software tools Unity 3D and Vuforia.
3. Augmented Reality (AR) positively influences the enhancement of learning quality, curriculum support, efficient evaluation, and technological advancement in the educational process.
4. Augmented Reality (AR) serves as an innovative instructional technology that facilitates creative cooperation among learners, enhances their creative thought abilities, and aids teachers in advancing forthcoming educational research.

The incorporation of mixed reality (MR) modern technology in educational institutions has gained traction as a promising tool for learning and learning. Its widespread adoption in current years has resulted in several benefits for learners, including enhanced motivation, problem-solving abilities, and overall learning experience. The utilization of Mixed Reality (MR) in education provides an excellent chance for learners to engage effectively in the education process. This is achieved by collaborative engagement and Solution-oriented analysis and resolution activities that involve the use of tangible things, such as robots, books, and maps, which are representative of real-world scenarios. The use of dynamic programs in the educational process enhances learners' focus and offers a more captivating and pleasurable education encounter compared to conventional methods (Banjar et al., 2023).

Hence, (MR) technology will enhance the existing digital educational landscape by offering learners engaging and pleasurable educational encounters. Furthermore, it is crucial to acknowledge that instructors make deliberate design choices based on instruction and the needs of students. They can utilize (MR) capabilities such as spatial visualization, interactive simulation, and virtual processing to align educational programs, instruction, and learner requires. This enables the creation of transforming instruction experiences that go beyond the superficial novelty of technology (Almufarreh, 2023).

2.5 The Quality of the Educational Process

Quality in education was established to address ongoing advancements in the field of learning and offers practical devices for informative institutions to meet the requirements, desires, and anticipations of their users both now and in later use. Quality is based on a shared commitment to consistently strive for excellence and prevent errors or shortcomings. It is not solely the responsibility of top management, but rather an obligation for all individuals within the educational institution to actively contribute to ongoing enhancement efforts. While endorsed by a quality structure, the quality of learning encompasses a couple significant aspects. The challenges can be effectively addressed by enhancing the quality of instruction, which encompasses both factors internal and external. Both the internal and external approaches to higher education are important, but the latter comes from outside the university and has a much larger impact on its success (Azizi et al., 2023).

The objective of incorporating cutting-edge learning and informational technologies in learning institutions is to equip learners with the skills to separately seek and explore information using contemporary technology. This includes the ability to study and evaluate information autonomously, generate logical findings, and assess the outcomes of their evaluation within the relevant setting. In the learning environment, the instructors establish the essential circumstances for the learner's growth, formation, learning, and instruction. Simultaneously, the instructor also fulfills administrative responsibilities within the college or university, guiding the learner toward creativity. In the context of learning technology, it refers to a systematic and interconnected series of actions carried out by the teacher. These actions are aimed at the deliberate and consistent implementation of a pre-determined learning process in practical settings or for resolving educational challenges (Mirzoxid et al., 2023).

2.6 Characteristics of the Educational Process Quality

Because it is tied to the goals and aims of educational institutions as well as the norms of a given system, program, or field of study, "quality" in education is an ever-evolving, multi-layered notion. According to research (Zongozzi, 2022) the following factors contribute to a high-quality learning environment:

a. The Exception Perspective: This dimension establishes a connection between quality and excellence, emphasizing that the primary objective of educational institutions is to cultivate the most exceptional pupils.
b. The Perfectionist Perspective: Quality is perceived as achieving constant and faultless outcomes.
c. Fit-for-purpose context: Quality is viewed as its ability to satisfy specific needs, that are defined by people (users of educational platforms). This is assessed based on the learning institutions' ability to carry out their mission statements or achieve the goals outlined in their learning applications.
d. The fit-for-purpose approach describes quality as the capacity of educating facilities to fulfill the demands of people, who are the users of educational services. The student sets the requirements, which are then evaluated according to the institution's capacity to fulfill its purpose statements or accomplish the objectives outlined in its educational modules.

Quality has a considerable impact on accountability in the higher education sector, since it redirects attention towards long-lasting enhancements. Consequently, the Ministry of Education and Training mandates that every educational institution must establish a dedicated quality assurance department or unit. This quality assurance process serves two primary objectives: ensuring accountability and facilitating continuous improvement, which can be synergistic. Modern quality assurance activities prioritize various aspects such as curriculum development, new education programs, qualification of lecturers, utilization of new technologies in higher education, provision of comprehensive knowledge and skills, accountability and responsibility, lifelong education, global citizenship, and vision (Ngoc et al., 2023).

The Electronic Quality Management System (EQMS) is a digital solution for managing and documenting quality management operations in a company. It facilitates quality control, compliance, and product development throughout the full value chain. The core of this system is the Digital Quality Management System (DQMS), which serves as a digital platform for managing quality. Digitization refers to the act of transferring an existing activity or process into a digital setting, while keeping its content and participants unchanged. This transition brings about alterations at four distinct levels, as identified by (Palladas, and Papakonstantinou, 2022):

a. Process level refers to the implementation of new digital instruments or the digitization of physical processes, resulting in changes in how work tasks and activities are performed.
b. Enterprise level (novel methods for delivering existing presentations, as well as novel presentations enabled by digitization).
c. Business area level (changing responsibilities and value chains).
d. Community level (changing forms of labor, enhancing effectiveness, and implementing digital infrastructure).

E-learning is a contemporary innovation that involves the incorporation of technology into the learning system. Through the advancement of Internet technology, there has been a significant transformation in communication and collaboration among academics. The digital revolution has facilitated unrestricted global accessibility to information, making it imperative to incorporate digital learning. The higher learning system aims to leverage Internet technology to compete and deliver high-quality education through digital alterations, technological innovations, and lightning-fast transformation. However, in spite of these efforts, there are several difficulties that hinder the educational process (Javed and Alenezi, 2023); (Danmuchikwali & Suleiman, 2020); (Cunha et al., 2020). These difficulties include:

a. Digital Literacy: All educators and learners require a fundamental comprehension of computers to effectively operate in a digital environment. Demonstrating digital fluency is the proficient utilization of technology to access knowledge, assess sources, generate.
b. Absence of direct teacher-student interactions: The extent of contact students have on campus, such as engaging in question and answer sessions, participating in discussions before and after instruction, attending office hours, and interacting in the classroom, is often underestimated. These opportunities for interaction are not available in digital education.
c. The necessity of self-discipline: Numerous students encounter challenges in exercising self-restraint within the higher education classroom setting, given the constant supervision of teachers and parents. In order to avoid forfeiting the benefits of social interaction by skipping classes or assignments, students must possess the ability and motivation to concentrate on the current task. In the realm of digital learning, it is convenient to bypass this task.

d. Technological obstacles: It is presumed that all individuals possess a cutting-edge laptop or desktop computer. Nevertheless, while being part of a generation that is highly reliant on digital technology, not every student has equitable access to the Internet, as the utilization of electronic gadgets has become widespread. The prevalence of mobile usage in online activities is a common practice among individuals. However, some persons face constraints in accessing the internet or Wi-Fi due to limitations in their smartphone plans.

It is anticipated that conventional quality processes will need to change in order to ensure the continued efficiency of learning organizations. Quality experts must adjust to a new period of developed technology and creativity. Thus, it is anticipated that experts and their learning organizations would enhance their practices and procedures concerning quality by integrating the most cutting-edge technology tools, media, and strategies. Student satisfaction with the educational process can be improved by the implementation of programming systems, which is the primary focus of the application of knowledge in university administration for quality assurance. Collecting, processing, evaluating, and maintaining the quality of information are all made easier with the help of modern information and communication technologies. The effective spread of information is also improved (Javed and Alenezi, 2023). The internet technologies are beneficial in higher learning as they assess the proficiency of both instructors and learners. They also help align teachers' competencies to empower students. Additionally, according to the European System for Digital Competency for educators, digital innovations can enhance educators' abilities on three different levels (Yan & Wei, 2022):

a. The learners' success is contingent upon the instructor's proficiency, as instructors must undergo training and adhere to guidelines. Failing to acquire adequate expertise in online technologies may hinder the learner's capability to navigate technology. The instructor's abilities and demeanor impact their effectiveness. The instructor's expertise and familiarity with the learning organization, as well as their fluency in advanced technologies and computer literacy, are indicative of their ability to foster inquiry-based education between learners.
b. The instructor exhibits proficiency in creating and executing educational programs, monitoring and assessing student performance and achieving the objectives of student centers.

Teachers' creativity, inspiration, confidence, impressions, and skills all play a role in how well they use online technology in the classroom. The university culture also plays a role in promoting cooperative learning in the digital field, management, structure improvement, and an accessible learning atmosphere.

3 Virtual Reality Technologies of Educational Process

Evaluating the quality of learning in higher learning is crucial, and studying this subject generally involves measuring learning quality as a comprehensive concept using self-reported data. Instructors' perceptions of educational quality may be subject to both overestimation and underestimation. However, when multiple variables of educational quality are combined to generate an overall measure, it effectively captures variations in educational quality across instructors (Daumiller et al., 2023). Consequently, the scholar examined numerous prior research and determined three primary aspects of the educational process's quality that align with the specific circumstances of this research.

The global higher education market has experienced significant disruptions due to recent advancements in digital technologies. Given these recent advancements colleges and universities must quickly and efficiently integrate technology into their courses, classrooms, and extracurricular to enhance the learning experience for their students. These difficulties have motivated professionals in higher learning to search for methods to meet rigorous certification standards and provide the best possible learning opportunities for their learners (Hanaysha et al., 2023).

Educators at all levels of the higher education system are required to gain unfamiliar information and cultivate digital skills, strategies, and techniques to effectively incorporate digital technology into the learning process and organization of education. This will enable students to utilize digital resources, including classroom technology, programs, games, and digital content/data (such as files, pictures, audio, and audiovisual), in a manner that enhances their knowledge advancement. The current scope of digital competencies encompasses expertise, mindset, and abilities in five distinct domains: 1. Data manipulation, 2. Communication and cooperation, 3. Group of digital content, 4. Security, 5. Troubleshooting (Roumbanis et al., 2023).

Recently, there has been a growing concentration on learning reforms due to modifications in the conventional learning curriculum. The traditional curriculum is defined by the transmission of information from teachers to students. With instructors being the focus of learning. However, a major drawback of this approach is the lack of emphasis on vital thinking and practical problem-solving skills. Consequently, learners thought is confined to predetermined trends. In order to address this issue, there has been a shift towards student-centered learning, which has led to changes in the traditional curriculum and the role of teachers. In this new approach, learners are seen as active participants in the educational process, while teachers provide knowledge and make necessary adjustments. As a result, there is a greater focus on communication and involvement among educators and learners (Abdigapbarova & Zhiyenbayeva, 2023; Atiyah and Zaidan, 2022).

Presently, educational institutions are equipping students for future employment opportunities that have not yet materialized. This entails imparting students with a set of prerequisites that are considered benchmarks of excellence and essential qualities for future triumph, such as the capacity to execute practical and significant assignments, and to employ acquired knowledge and abilities in real-world scenarios. The following are the important talents for the twenty-first century that enable students to make valuable contributions to the global workforce (Loureiro & Gomes, 2023).

Improving learner involvement and motivation in learning has become a prominent concentration in higher education learning practices. This is particularly evident in the expanding usage of cooperative e-learning and the establishment of virtual education societies. Online learning emphasizes cooperative learning, knowledge-building, and the use of Internet programs to transform both formal and non-formal education. Studies indicate that in the knowledge age, the combination of an intelligent and adaptable education platform, together with carefully crafted team-based activities in higher education institutions, can enhance student involvement in a wide range of educational activities and settings (Kang & Zhang, 2023).

Therefore, technology can assist pupils in acquiring knowledge and enhancing their skills and levels of accomplishment. Additionally, it results in a reallocation of educational resources from prioritizing the instructor to prioritizing the student. Furthermore, it offers students many chances for extracurricular education, encompassing a wide range of programs, sites, videos, online lectures, e-books, and more. In addition, digital technologies are generally acknowledged as highly effective devices for advancing education by allowing teachers to play the role of organizers inside the classroom setting. Consequently, learners should assume responsibility for their own education and actively participate in the educational environment and methods, as well as evaluate their own advancement (Pratiwi & Waluyo, 2023).

The disruptions in higher education have sparked concerns regarding the optimal curriculum design for online education, ensuring the inclusion of all students. Therefore, a holistic approach to education design, combined with curriculum research models, offers a promising method to enhance student learning and foster engagement in online education. The online curriculum seeks to enhance the learning, and perceived contentment of all students. Additionally, it strives to improve the caliber of online education by making instruction more efficient. Education practices encompass components that are introspective, instructive, and communal in nature. An intriguing finding that arose is the significance of the collaboration between the designer, academic, and instructional designer in the process of curriculum development (Sheidan and Gigliotti, 2023). To create a comprehensive educational program, it is essential to incorporate technology into the educational program. Technology has significantly influenced education, particularly in curriculum design. It can be utilized to enhance students' acquisition of skills and promote their autonomy. Therefore, the syllabus should be regularly updated to align with the rapidly evolving world, while considering the integration of technology (Tasya, 2023).

In order to enhance clarity of the vision, one of the technologies employed throughout the curriculum is Artificial Intelligence (AI). The overarching objective is to cultivate a skilled workforce that encompasses the essential proficiencies demanded by the twenty-first-century job market and the global governmental requirements, utilizing AI as a central component. To meet the demands of the 21st century, competent human resources are essential, and educational institutions are positioning themselves to take the lead in meeting this universal social need. When implementing (Al) throughout the curriculum, every student is offered a variety of (Al) possibilities and is motivated to engage. Educational institutions are utilizing substantial investments in campus-wide artificial intelligence (AI) to develop innovative curricula and implement activities that promote interdisciplinary interaction, all while ensuring learners are prepared for their future careers (Southworth et al., 2023).

4 Methodology

The educational applications of Metaverses are explored using a descriptive analytic methodology in this work. Methodical and open, a systematic literature review identifies relevant research, assesses their quality, and synthesizes their methods and findings.

Here, it means taking a look back at research that has been done on the topic of using Metaverse methods in educational settings (Singh and Thurman, 2019). Web of Science and Scopus, two databases known for their extensive coverage and high impact factors, were searched using the provided terms. This study limits its literature evaluation to recent case studies that concentrate solely on the use of the Metaverse in educational settings because of its rising popularity in this area. The featured case studies were thoroughly reviewed after being retrieved from databases. Studies were chosen based on their quality and relevance, as well as a set of preset inclusion and exclusion criteria. In order to conduct this research, previous case studies were considered to provide a foundation for identifying the keywords. The keyword "Metaverse" was utilized and paired with keywords linked to the educational process quality.

Salloum et al. (2023) evaluate how introducing Metaverse technology affected Oman Arab University. The process required the construction of a theoretical framework that takes into account novelty, awareness of context, enjoyment, ease of use, prevalence, challenge, and worth. Nine hundred fifty-three individuals participated in the online poll that yielded the collected data. The results indicate that a forward-thinking school environment may influence how teachers and students see technology. Understanding how the Metaverse functions as a modern educational resource might influence students' attitudes about the introduction of new technologies and help universities craft policies that enhance students' learning. According to the results, fresh inventions have a significant impact on whether or not people adopt new technologies. According to the findings, creativity is crucial to Metaverse's success. The widespread availability of Metaverse, however, does not appear to significantly increase its utilization. Metaverse adoption is also heavily impacted by aspects including understanding of context, perceived challenge, and enjoyment. Researchers have learned, through immersion in the Metaverse, how university administrators can most effectively raise awareness of the Metaverse among faculty and students. The benefits of Metaverse can be better communicated to educators if conferences and seminars are held, financial incentives are offered to teachers who adopt the platform, and subject matter experts are invited to speak on the topic. Khalil et al. (2023) used a survey and focus groups; it investigates whether or not university educators in Pakistan are using and/or accepting of the Metaverse in the classroom. Six factors—effort expectation (EE), performance expectation (PE), social influence (SI), facilitating conditions (FC), behavioral intentions (BI), and use behavior)—make up the UTAUT model, a complete paradigm for accepting new technologies.

The research used a mixed-methods strategy, analyzing the results of the surveys and interviews with both quantitative and qualitative techniques. The research sampled (315) students and (10) faculty members from four Lahore colleges. The findings of the study showed that both educators and students were open to the idea of using Metaverse in the classroom. Therefore, the study recommended that university administration encourage Metaverse use in the classroom by strengthening infrastructure, scheduling training for faculty, and providing technical support for students and instructors. Siswantoro et al. (2023) confirmed VR recreations of ship bridges give students the chance to hone their practical abilities in a safe, controlled environment before using them in the real world. As an example, we looked into the Metaverse zone of Tanjung Priok Port. Blender was used to digitize physical objects into 3D assets, and then Unity was used to assemble a virtual reality application bundle. The virtual ship in the Metaverse can be navigated both forward and backward in time. The app's success can be measured by the favorable feedback it has received from its 30 or so student users. A score of 908 out of 1050 indicates that this VR program is of the highest quality.

5 Conclusions

Finding a rising role for Metaverse technologies in the medical, engineering, and scientific disciplines, this study provides a descriptive analysis of the topic of metaverse and education. Metaverse technologies are becoming increasingly popular in the classroom. Due to a lack of investment in the necessary technological infrastructure, there is resistance and rejection of the expanding function of these technologies. In addition, in this context, "Metaverse technologies" refer to those that bridge the gap between knowledge systems and ICT, and they have run into a number of roadblocks, most notably the absence of a hospitable organizational culture in some developing nations. As a result, here are some things to keep in mind: First, raising awareness in the academic community about the value of using Metaverse technologies. Second, supporting research and development of Metaverse technologies by establishing a foundation for knowledge management and information and communication technology. Finally, integrating Metaverse tools into hybrid courses. Fourth, including students' use of Metaverse technologies into the grading of their academic work. Fifth, make connections with schools that are already using Metaverse technologies to form twinning and collaboration programs. The usage of metaverse technology in higher education is important due to its successful function in optimizing the learning process for both students and teachers. It is imperative for higher education institutions to embrace Metaverse technology in order to optimize the quality of university education.

References

Singh, V., Thurman, A.: How many ways can we define online learning? A systematic literature review of definitions of online learning (1988–2018). Am. J. Distance Educ. **33**(4), 289–306 (2019)

Abdigapbarova, U., Zhiyenbayeva, N.: Organization of student-centered learning within the professional training of a future teacher in a digital environment. Educ. Inf. Technol. **28**(1), 647–661 (2023)

Al-Abrrow, H., Fayez, A.S., Abdullah, H., Khaw, K.W., Alnoor, A., Rexhepi, G.: Effect of open-mindedness and humble behavior on innovation: mediator role of learning. Int. J. Emerg. Mark. (2021)

Al-Adwan, A.S., Li, N., Al-Adwan, A., Abbasi, G.A., Albelbisi, N.A., Habibi, A.: Extending the technology acceptance model (TAM) to predict university students' intentions to use metaverse-based learning platforms. Educ. Inf. Technol. 1–33 (2023)

Albahri, A.S., et al.: Based on the multi-assessment model: towards a new context of combining the artificial neural network and structural equation modelling: a review. Chaos, Solitons Fractals **153**, 111445 (2021)

Alfaisal, R., Hashim, H., Azizan, U.H.: Metaverse system adoption in education: a systematic literature review. J. Comput. Educ. 1–45 (2022)

AL-Fatlawey, M.H., Brias, A.K., Atiyah, A.G.: The role of strategic behavior in achievement the organizational excellence “analytical research of the manager’s views of Ur State Company at Thi-Qar Governorate”. J. Admin. Econ. **10**(37) (2021)
Almufarreh, A.: Exploring the potential of mixed reality in enhancing student learning experience and academic performance: an empirical study. Systems **11**(6), 1–18 (2023)
Alnoor, A., et al.: How positive and negative electronic word of mouth (eWOM) affects customers’ intention to use social commerce? A dual-stage multi group-SEM and ANN analysis. Int. J. Hum.–Comput. Interact. 1–30 (2022)
Alsalem, M.A., et al.: Rise of multiattribute decision-making in combating COVID-19: a systematic review of the state-of-the-art literature. Int. J. Intell. Syst. **37**(6), 3514–3624 (2022)
Atiyah, A.G.: The effect of the dimensions of strategic change on organizational performance level. PalArch’s J. Archaeol. Egypt/Egyptol. **17**(8), 1269–1282 (2020)
Atiyah, A.G., Zaidan, R.A.: Barriers to using social commerce. In: Alnoor, A., Wah, K.K., Hassan, A. (eds.) Artificial Neural Networks and Structural Equation Modeling, pp. 115–130. Springer, Singapore (2022). https://doi.org/10.1007/978-981-19-6509-8
Azizi, M.H., Bakri, S., Choiriyah, S.: Implementation of total quality management in the ministry of religion-based education. Nidhomul Haq: Jurnal Manajemen Pendidikan Islam **8**(1), 125–136 (2023)
Banjar, A., Xu, X., Iqbal, M.Z., Campbell, A.: A systematic review of the experimental studies on the effectiveness of mixed reality in higher education between 2017 and 2021. Comput. Educ. X Real. **3**(100034), 1–15 (2023)
Contreras, G.S., González, A.H., Fernández, M.I.S., Martínez, C.B., Cepa, J., Escobar, Z.: The importance of the application of the metaverse in education. Mod. Appl. Sci. **16**(3), 34–40 (2022)
Cunha, M.N., Chuchu, T., Maziriri, E.: Threats, challenges, and opportunities for open universities and massive online open courses in the digital revolution. Int. J. Emerg. Technol. Learn. (iJET) **15**(12), 191–204 (2020)
Danmuchikwali, B.G., Suleiman, M.M.: Digital education: opportunities, threats, and challenges. Jurnal Evaluasi Pendidikan **11**(2), 78–83 (2020)
Daumiller, M., Janke, S., Hein, J., Rinas, R., Dickhäuser, O., Dresel, M.: Teaching quality in higher education: agreement between teacher self-reports and student evaluations. Eur. J. Psychol. Assess. **39**(3), 1–6 (2023)
Elbyaly, M.Y.H., Elfeky, A.I.M.: The effectiveness of a program based on augmented reality on enhancing the skills of solving complex problems among students of the Optimal Investment Diploma. Ann. For. Res. **66**(1), 1569–1583 (2023)
Fadhil, S.S., Ismail, R., Alnoor, A.: The influence of soft skills on employability: a case study on technology industry sector in Malaysia. Interdiscip. J. Inf. Knowl. Manag. **16**, 255 (2021)
Fitria, T.N., Simbolon, N.E.: Possibility of metaverse in education: opportunity and threat. SOSMANIORA: Jurnal Ilmu Sosial dan Humaniora **1**(3), 365–375 (2022)
Gatea, A.A., Marina, V.: Higher education funding in Iraq in terms of the experience of particular developed countries. Int. J. Adv. Stud. **6**(1), 8–17 (2016)
Hamid, R.A., et al.: How smart is e-tourism? A systematic review of smart tourism recommendation system applying data management. Comput. Sci. Rev. **39**, 100337 (2021)
Hanaysha, J.R., Shriedeh, F.B., In’airat, M.: Impact of classroom environment, teacher competency, information and communication technology resources, and university facilities on student engagement and academic performance. Int. J. Inf. Manag. Data Insights **3**(2), 1–12 (2023)
Hassanzadeh, M.: Metaverse, metaversity, and the future of higher education. Sci. Tech. Inf. Manag. **8**(2), 7–22 (2022)
Hidajat, F.A.: Augmented reality applications for mathematical creativity: a systematic review. J. Comput. Educ. 1–50 (2023)

Hwang, G.J., Chien, S.Y.: Definition, roles, and potential research issues of the metaverse in education: an artificial intelligence perspective. Comput. Educ. Artif. Intell. **3**(100082), 1–6 (2022)

Javed, Y., Alenezi, M.: A case study on sustainable quality assurance in higher education. Sustainability **15**(10), 1–14 (2023)

Kaddoura, S., Al Husseiny, F.: The rising trend of Metaverse in education: challenges, opportunities, and ethical considerations. PeerJ Comput. Sci. **9**, e1252, 1–33 (2023)

Kang, X., Zhang, W.: An experimental case study on forum-based online teaching to improve student's engagement and motivation in higher education. Interact. Learn. Environ. **31**(2), 1–13 (2023)

Khalil, A., Haqdad, A., Sultana, N.: Educational metaverse for teaching and learning in higher education of Pakistan. J. Positive Sch. Psychol. 1183–1198 (2023)

Khaw, K.W., et al.: Modelling and evaluating trust in mobile commerce: a hybrid three stage Fuzzy Delphi, structural equation modeling, and neural network approach. Int. J. Hum.-Comput. Interact. **38**(16), 1529–1545 (2022)

Liberman, L., Dubovi, I.: The effect of the modality principle to support learning with virtual reality: an eye-tracking and electrodermal activity study (2023)

Lin, H., Wan, S., Gan, W., Chen, J., Chao, H.C.: Metaverse in education: vision, opportunities, and challenges. In: 2022 IEEE International Conference on Big Data (Big Data), pp. 2857–2866. IEEE, December 2022

Loureiro, P., Gomes, M.J.: Online peer assessment for learning: findings from higher education students. Educ. Sci. **13**(3), 1–19 (2023)

Manhal, M., Al-khalidi, A., Hamad, Z.: Strategic network: managerial myopia point of view. Manag. Sci. Lett. **13**(3), 211–218 (2023)

Marks, B., Thomas, J.: Adoption of virtual reality technology in higher education: an evaluation of five teaching semesters in a purpose-designed laboratory. Educ. Inf. Technol. **27**(1), 1287–1305 (2022)

Mirzoxid, B., Onaxon, B., Mexribonu, A.: Use of pedagogical technologies in improving the quality of education. Новости образования: исследование в XXI веке **1**(6), 1013–1018 (2023)

Ngoc, N.M., Hieu, V.M., Tien, N.H.: Impact of accreditation policy on quality assurance activities of public and private universities in Vietnam. Int. J. Publ. Sector Perform. Manag. 1–15 (2023)

Onecha, B., Cornadó, C., Morros, J., Pons, O.: New approach to design and assess metaverse environments for improving learning processes in higher education: the case of architectural construction and rehabilitation. Buildings **13**(5), 1–21 (2023)

Onggirawan, C.A., Kho, J.M., Kartiwa, A.P., Gunawan, A.A.: Systematic literature review: the adaptation of distance learning process during the COVID-19 pandemic using virtual educational spaces in metaverse. Procedia Comput. Sci. **216**, 274–283 (2023)

Palladas, A., Papakonstantinou, N.: Advantages and disadvantages of a digital quality management system. Email: quality20zavvar@ gmail. com, 25 (2022)

Pratiwi, D.I., Waluyo, B.: Autonomous learning and the use of digital technologies in online English classrooms in higher education. Contemp. Educ. Technol. **15**(2), 1–16 (2023). Sheridan, L., & Gigliotti, A. (2023). Designing online teaching curriculum to optimise learning for all students in higher education. The Curriculum Journal, 1–23

Sheridan, L., Gigliotti, A.: Designing online teaching curriculum to optimise learning for all students in higher education. Curriculum J. 1–23 (2023)

Recker, J., Lukyanenko, R., Jabbari, M., Samuel, B.M., Castellanos, A.: From representation to mediation: a new agenda for conceptual modeling research in a digital world. MIS Q. (2021)

Roumbanis Viberg, A., Forslund Frykedal, K., Sofkova Hashemi, S.: The teacher educator's perceptions of professional agency–a paradox of enabling and hindering digital professional development in higher education. Educ. Inq. **14**(2), 213–230 (2023)

Salloum, S., et al.: Sustainability model for the continuous intention to use metaverse technology in higher education: a case study from Oman. Sustainability **15**(6), 5257, 1–19 (2023)

Siswantoro, N., Haryanto, D., Hikmahwan, M.A., Pitana, T.: Implementation of metaverse for modeling the ship bridge simulator based on virtual reality as educational purposes, case study: Tanjung Priok Port area. In: IOP Conference Series: Earth and Environmental Science, vol. 1166, no. 1, p. 012053. IOP Publishing, May 2023

Smutny, P.: Learning with virtual reality: a market analysis of educational and training applications. Interact. Learn. Environ. 1–14 (2022)

Southworth, J., et al.: Developing a model for AI across the curriculum: transforming the higher education landscape via innovation in AI literacy. Comput. Educ. Artif. Intell. **4**, 1–10 (2023)

Tasya, D.H.: The role of technology in English curriculum design. PUSTAKA: Jurnal Bahasa dan Pendidikan **3**(4), 42–57 (2023)

Tlili, A., et al.: Is Metaverse in education a blessing or a curse: a combined content and bibliometric analysis. Smart Learn. Environ. **9**(1), 1–31 (2022)

Yan, W., Wei, C.: Efficiency of Digital Technology Use in Schools. Introduction to Research Methods RSCH 202, pp. 1-17 (2022)

Zongozzi, J.N.: Accessible quality higher education for students with disabilities in a South African open distance and e- learning institution: challenges. Int. J. Disabil. Dev. Educ. **69**(5), 1645–1657 (2022)

Influence of Authentic Leadership Practices on Innovative Work Behaviour in Higher Educational Institutions: A Virtual Reality Perspective

Hafiza Saadia Sharif[1], Al-Amin Bin Mydin[1(✉)], and Hussain A. Younis[2,3]

[1] School of Educational Studies, University Sains Malaysia, 11800 Gelugor, Penang, Malaysia
saadiasharif@lgu.edu.pk, alamin@usm.my

[2] College of Education for Women, University of Basrah, 61004 Basrah, Iraq

[3] School of Computer Sciences, Universiti Sains Malaysia, 11800 Gelugor, Penang, Malaysia

Abstract. Virtual reality is an emerging technology that has created immersive experiences and captured the interest of users and industry in various disciplines in the 21st century. The little research and scarcity reported in the literature in the field of using virtual reality technologies to stimulate creative behavior is considered a fundamental issue that needs urgent intervention. The current study intends to examine how four authentic leadership dimensions influence authentic leadership on innovative work behaviors through the moderating role of metaverse technology in higher educational institutions. Innovative work behavior at institutions of higher education has emerged as a difficult problem, according to a recent study in this area. Permanent faculty members from four private institutions of higher education located in Pakistan's Lahore city constitute the study's population. To collect information from higher education institution faculty, a survey study design was used. Ten percent of the entire population was selected as the sample size using a proportionate convinces based sampling method. The respondents' attitudes on two topics were assessed using the authentic leadership scale and the innovative work behavior scale, two measuring scales. On a five-point Likert scale, which was used to gauge the respondents' thoughts? Through the evaluation of experts and the Coronach alpha value the questionnaire's reliability and validity were examined in a pilot study. The Cranach alpha value for AL was 0.813 and for IWB was 0.817. Using SPSS version 25, a multiple linear regression technique was used to analyze the data. According to the four dimensions' the behaviors of "internalized moral perspective" and "relational transparency" have a statistically significant influence on faculty members' innovative work behavior in higher educational institutions, However, "self-awareness" and "balanced processing" have negative effects. To obtain a thorough and more universal understanding of the subject, instructors at higher educational institutions should conduct their studies across different parts of Pakistan. As far as the researchers are aware. This is the first study on faculty members' innovative work behaviors and the self-awareness of authentic leadership at higher educational institutions in Pakistan.

Keywords: Authentic leadership · Innovative work behavior · Virtual reality · Higher educational Institutions

M. Al-Emran et al. (Eds.): IMDC-IST 2024, LNNS 876, pp. 175–187, 2023.
https://doi.org/10.1007/978-3-031-51300-8_12

1 Introduction

Virtual reality is considered a substitute for real reality, using a number of programs and platforms that increase interaction between people. The use of virtual reality technologies has increased by many people, and immersive experiences of virtual reality technologies have spread. Virtual reality applications are being used by many business models (Alfaisal et al., 2022). The popularity of virtual reality applications, such as the metaverse, has increased due to their advanced ability to provide superior technological services and experiences. Metaverse technologies have combined new experiences in many areas for users, such as education, health, sports, business, production processes, and others (Gai et al., 2023). The ease of using metaverse technologies and their technical ability to provide high-quality displays that allow the integration of virtual reality with real reality have made them more widely used. Investments in virtual reality technologies increased and reached $6.1 billion in 2020. The market for investments in metaverse technologies will reach $20.9 billion in 2025 (Dincelli & Yayla, 2022). Companies and educational institutions have focused on applying and using virtual reality technologies represented by the metaverse to increase user satisfaction and provide superior services (Xi et al., 2023). The virtual reality industry has changed the technological landscape and national strategy of many countries. Scientists have contributed to the development of metaverse technology and have shown tremendous interest. The literature has confirmed that metaverse technology in business models is the fundamental driver of creative user behavior (Pamucar et al., 2023). The higher educational institutions dynamic is a worldwide occurrence that is equally present in developing nations in the vein of Pakistan. Higher educational Institutions that instruct students in a variety of circles of life, are measured as the maximum sources of knowledge and awareness creation, according to Carter (Carter et al., 2023). Faculty personnel whose primary responsibilities are teaching, conducting research, or taking part in extracurricular activities at their institutes make up the academic staff. They are essential resources for any educational programs and institution's success. Therefore, academics' innovative work behaviors are crucial to the academic achievement of higher education institutions (Ефимова & Латышев, 2023). Innovative work behavior shows different attitudes towards their work circumstances and is a condition of faculty members' favorable attitudes towards their employment. Additionally, inventive work has been identified as the key component among faculty members (Saxena & Prasad, 2023).

Authentic leadership has a significant influence on employees' innovative work behavior; according to past studies, authentic leadership empowers and uplifts their staff by remaining truthful, open, and supportive of them. These qualities have been shown to be significant indicators of employee voice behavior. In particular, authentic leadership encourages openness in communications with employees, maintains high standards of ethical behavior, and clarifies data in an objective manner (Carvalho et al., 2023). By exhibiting these traits, leaders can increase followers' sense of safety and psychological support, which in turn encourages followers to take sensible risks (Kaya & Karatepe, 2020). This commitment empowers workers to express their unconventional thoughts and any point of view without fear. Mentioned the previously highlighted employee voice behavior (Teng & Yi, 2022).

Previous studies have supported the link between authentic leadership and innovative work behavior (Mubarak et al., 2021). Analyze the association between authentic leadership and innovative work behavior since it has been shown to have a positive impact on employee innovation in the workplace (Işık et al., 2021). Additionally, every organization must adopt risk-taking behavior in order to remain viable in the unpredictable and rapidly evolving business climate (Faulks et al., 2021).The connection between authentic leadership and innovative work behavior has received less attention in the past. In the setting of higher education institutions (Yamak & Eyupoglu, 2021). Encountered the significant influence of authentic leadership on innovative work behavior. In a comparable manner, it was claimed (Phuong & Takahashi, 2021). In the competitive and dynamic world of today's environmental conditions, innovation is crucial for the success of organizations ((Kleynhans et al., 2021). According to the professionals, leadership is one of the most crucial elements that affects both creativity and innovativeness (Grošelj et al., 2021).The rising academic attraction with innovative settings is also shown in authentic leadership, which has grown into a "widespread, developing social trend" and a "golden standard for leadership (Durrah & Kahwaji, 2022). The attention of professionals is increasing, as is the curiosity among academics. Along with education, innovation in the workplace is essential for workers, organizations, and societies. As such, it plays a significant role in the European Union's objectives for globalization and knowledge-based nations. At the organizational and national levels, innovation is an essential component of social and economic development, while at the individual level, innovation at work is a requirement for greater innovative work behavior (Geltzer, 2017).

Thus, the research that has already been done enables us to begin to comprehend the factors that affect employees' innovative work behavior. There have been numerous studies on both leadership and innovativeness, but few have examined the boundary conditions of the relationships between authentic leadership and innovative work behavior (Sethibe & Steyn, 2017).We are seeing varying empirical evidence on the relationship between authentic leadership and innovative work behavior, despite the research findings supporting the positive influence authentic leadership has on encouraging innovative work behavior. a multi-dimensional framework of organizational innovation: a systematic review of the literature (Mishra et al., 2019). Mostly insignificant information is known about the related boundary components that influence leaders' capacities to foster innovation in companies. Therefore; this is a good time to consider the criteria for authenticity and transformative leadership in cutting-edge workplaces. The broad theoretical framework of innovative knowledge environments serves as our guide (Isaksen, 2017). It contends that individuals who engage in innovative activities (for example, by working in innovative environments, such as research and development are better able to comprehend this phenomenon (Latynina et al., 2020).

Whereas variables at the foundations of an organization are affected by circumstances at the top levels, they are embedded in numerous different organizational levels of influence in addition, surroundings with attributes that beneficially impact people working creatively to generate new information or ideas, whether they do it alone, in teams, within a single organization, or in partnership with others, are important. This paper's goal is to contribute to the growing body of research literature on the interplay between authentic leadership and the promotion of innovative work behavior. Authentic leadership in motivating innovative work behavior, captivating into explanation the relationship in motivating innovative work behavior (Grošelj et al., 2021).

Authentic leadership is outstanding for promoting innovative thinking and creativity considering its characteristics (Oh & Oh, 2017). An attribute of authentic leadership is the ability to learn from mistakes and use creativity and knowledge to advance their followers (Allen-Ile et al., 2020). Leaders who are more self-aware value and encourage inventive and creative behavior. More so than less authentic leaders, highly authentic leaders inspire innovative work behavior in their followers. The followers get motivated and excited when they believe their leader to be an authentic leader (Oh & Oh, 2017). Such supporters are more likely to grow novel ideas because they are more self-aware in their ability to apply novel ideas and are better able to deal with challenges and opportunities (Xu et al., 2017). Increase their followers' positive psychological capital, including their optimism, self-confidence, and adaptability (Novitasari et al., 2020). Their adherents are less fearful of rejection or failure, and they are more willing to attempt new things (Ribeiro et al., 2018). Even if the innovation fails, followers who believe their workplace encourages experimentation are more likely to continue (Sengupta et al., 2023). The major objective of this study is to determine how four authentic leadership dimensions influence faculty members' innovative work behavior through the moderating role of metaverse in private higher educational institutions.

1. To investigate how "Self-Awareness" influence faculty members' innovative work behaviour in higher educational institutions.
2. To investigate the influence of "Internalised Moral Perspective" on faculty members' innovative work behaviour in higher educational institutions.
3. To investigate how "Balance Processing" influence faculty members' innovative work behaviour in higher educational institutions.
4. To investigate the influence of "Relational Transparency" on faculty members' innovative work behavior in higher educational institutions.
5. To investigate the moderating role of "metaverse" in higher educational institutions.

2 Literature Review and Hypotheses Development

2.1 Authentic Leadership

It has been proposed that authentic leadership is a pattern of behavior that draws upon and promotes positive psychological capacities and a positive ethical climate in order to further promote more self-awareness, an internalized moral perspective, balanced information processing, and relational transparency on the part of leaders while working with followers. This encourages followers to grow professionally. The study of authentic leadership has grown in popularity in the field of Higher educational organizations behavior during the past ten years (Farid et al., 2020).

Authentic leaders encourage and motivate followers to accomplish goals (Crawford et al., 2020a, b). They do this with the help of increased awareness and effective communication. According to recent studies, authentic leadership may benefit organizations because it has been connected to a variety of outcomes, including individual creativity, performance, customer orientation, and employee retention (Duarte et al., 2021). An authentic leader can increase innovative work behavior such as commitment, passion, sense of duty, contentment, and immersion to demonstrate better action for the growth of the organization and the welfare of the workforce (Kiersch & Peters, 2017).Authentic leadership embodies inspiration, creativity, and effective communication. He or she is self-developed and has a visionary spirit (Crawford et al., 2020a, b). Self-improvement is crucial for maintaining employee motivation. Authentic leaders consistently display both emotion and reason. They are always aware of the importance of their coworkers. They are adept at managing successfully (Bandura et al., 2019). Authentic leaders develop their teams, create open lines of communication with their employees, and always consider their wellbeing. They are aware that everyone should be treated with respect. Authentic leaders are self-aware, morally upright, and exhibit understanding of their own principles and beliefs. The initial element essential to the overall growth of an organization is for followers and leaders to engage positively. This form of connection relies heavily on confidence. The development of an organization is greatly influenced by a leader's self-assurance, bravery, helpfulness to subordinates, openness, and inventiveness Efficiency is seen as a feature of true leaders; authentic leaders are people who act as leaders by strongly encouraging their supporters to pursue higher employment (Saeed & Ali, 2019). An authentic leader is one who offers himself, declines to offer him, but never lends himself. The highest level of self-sacrifice is always expected of authentic leaders, and they are always built for modesty and self-effacement. Although authentic leaders can exhibit considerable self-sacrifice in the service of others, he never compromises his moral standards or permits him/her to be taken advantage of. At all costs, authentic leaders never waver from their beliefs and ideas. Authentic leaders' contempt the well-traveled path and is a surpassing intelligence with an exceptional creative personality (Azanza et al., 2013).

2.2 Innovative Work Behavior

A variety of definitions for innovative work conduct have been offered by researchers. Innovation in the workplace is "the production or adoption of useful ideas and idea implementation, which begins with problem recognition and the generation of ideas or solutions (Hashim, 2021). It is "the deliberate generation, promotion, and realization of new ideas within a work role, work group, or organization," according to Janssen (2000). In a similar vein, innovative work behavior is a process that should improve the efficiency and effectiveness of problem-solving in the workplace (Knezović & Drkić, 2021). According to a more current definition provided (Alnajjar & Hashim, 2020). It is "a set of repeated actions that are initiated by individuals and executed by groups within organizations based on continuous needs for improvement so that individuals, groups, and organizations would benefit" (Alhmoudi et al., 2022). Innovative work behavior encompasses several stages. Janssen (2000) recognized three stages of innovative work behavior: concept genesis, promotion, and realization. To enable innovation, employees must come up with fresh ideas or solutions to issues. This is known as idea generation. Building alliances, gaining support for the invention by outlining its benefits and success to prospective allies, and locating sponsors are all considered aspects of idea promotion (Akram et al., 2016). The creation of a sample or model of a novel product, system, or procedure that may be used as a component of standard procedures is also necessary for idea realization (De Jong & Den Hartog, 2010). According to earlier studies, there is a significant connection between innovative work behavior and a number of outcomes, including employee performance commitment (Hashim, 2021).

2.3 Metaverse

In recent years, the concept of the "Metaverse" has exploded in popularity, attracting the attention of techies and business moguls alike. It represents a virtual, interconnected universe where individuals interact, work, play, and explore, using digital avatars and real-time simulations. Often touted as the next evolution of the internet, the Metaverse is a concept deeply rooted in science fiction, but it's quickly becoming a reality. The Metaverse is not a single entity or platform but rather a convergence of various technologies, including augmented reality (AR), virtual reality (VR), blockchain, and artificial intelligence (AI) (Hudson-Smith & Batty, 2022). The ultimate goal is to make a digital experience so smooth and compelling that it makes us forget we're in the real world. It offers the promise of a boundless, persistent digital space where users can engage in a multitude of activities, from socializing to working, shopping to attending events. The potential for the Metaverse to revolutionize interpersonal relationships is one of its defining features. Combining virtual reality and augmented reality allows users to build unique avatars and interact with others in ways that are nearly identical to the real thing (Maheswari et al., 2022). Imagine attending a virtual concert, where you're not just watching a livestream but also dancing and chatting with friends as if you were in the same physical space. This has profound implications for communication, collaboration, and entertainment. There are also profound ramifications for business on the Metaverse. It has the potential to completely alter the way we do business and hold meetings. Already, businesses are investigating the potential of online meeting rooms and workplaces to improve the communication and collaboration of distributed teams (Gokasar et al., 2023). Online shopping stands to gain as well, with virtual showrooms and experiences reshaping the way we shop from the comfort of our own homes. Blockchain technology's ability to record digital transactions and transfer of ownership is fundamental to the functioning of the Metaverse. Users can buy, sell, and trade digital real estate, virtual items, and even land. As a representation of this ownership in the Metaverse, NFTs (Non-Fungible Tokens) make it possible for users to prove they are the rightful owners of everything from digital artwork to virtual real estate. However, the development of the Metaverse also raises concerns. Privacy, security, and digital identity are paramount issues that need to be addressed. Strong data protection and identity verification are becoming increasingly important as more of our lives move online. Furthermore, it is important to guarantee access and equal participation in the Metaverse so that it does not exacerbate preexisting inequities (Njoku et al., 2023). Thus, we assume that:

1. "Self-Awareness" has a statistically significant influence on faculty members' innovative work behaviours in higher educational institutions.
2. The "Internalized Moral Perspective" has a statistically significant influence on faculty members' innovative work behaviours in higher educational institutions.
3. "Balance Processing" has a statistically significant influence on faculty members' innovative work behaviours in higher education institutions.

4. There is a statistically significant correlation between "Relational Transparency" and faculty members' innovative work behavior.
5. The relationship between authentic leadership and innovative work behavior will be stronger when metaverse is high.

3 Methodology

3.1 Research Design

To evaluate and describe a population's characteristics, such as perceptions, attitudes, or behaviors, the researcher can conduct a survey of a selected group of individuals from the sample or the entire population. After using a quantitative research technique in which all study variables are measured simultaneously throughout the same time, the researcher employs a cross-sectional research technique (Rahi, 2017). The goal of quantitative research is to identify the sample population, which is a particular group of people. To address the research issues, quantitative research relies on data that is measured or observed (Apuke, 2017).

In comparison to a longitudinal approach, which collects data from respondents' multiple times at various time intervals, a cross-sectional approach collects data from respondents only once and yields results relatively quickly the longitudinal approach is very useful for examining the dynamics of the research over a range of time periods. The benefits of survey research enable large amounts of data to be collected quickly and simply Scales for measuring and research instruments. To gather data for the current investigation, a survey instrument with structured but typical questions was created. To collect enough information from the target population, the questionnaire for the current study has been modified from previous research. The authentic leadership scale (AL) and the innovative work behaviour scale (IWB) was two measuring scales that were modified to create the questionnaire acknowledged that using a questionnaire is the quickest as well as most accurate method to get data from a large, representative sample. The self-administered survey for this study has 33 items in two portions, as indicated below (Table 1).

Table 1. Section of Research Instrument.

Sections	Variables	Dimensions	Items	Source
A	Authentic Leadership	Self-Awareness Internalized Moral Perspective Balance Processing Rational transparency	16	(e.g. Janssen, 2000; Kleysen & Street, 2001; Scott & Bruce, 1994)
B	Innovative work behaviour	Idea Exploration Idea Generation Idea Chomping Idea Implementation	17	Created by Walumbwa and associates

3.2 Sample Size

The population of Lahore, Pakistan's private higher education institutions is made up of their faculty staff members of the current study. The chosen demographic comprises of academic staff members who are between the ages of 24 and 42, have a Ph.D. and M.Phil. You have at least a year's worth of classroom instruction experience in a private higher education setting. All full-time teaching faculty members, or academics whose exclusive focus is on teaching in private higher education institutions in Lahore, are the study's target group.

Since none of their departments had any additional part-time administration positions. It was because Lahore, Pakistan, has so many private higher education institutions. The capital of Pakistan's "Punjab" state, Lahore, is home to 22 private higher educational institutions. Lahore is renowned for its diverse academic environment and is also known as "the city of colleges" due to the abundance of colleges, universities, and institutions that grant degrees. To pursue their academic goals, a diversified multicultural academic community of students, teachers, and researchers travel to Lahore city from all over Pakistan. High multicultural diversity in the academic community, the current study was concentrated in the city of Lahore. The public entity was not included in the current analysis because it only looked at private higher education institutions in Lahore. Only four private higher educational institutions colleges were chosen at the researcher's convenience out of the 22 private higher educational institutions located in Lahore city, keeping in mind the time limit, logistics, and resources.

4 Data Analysis

The sample size for the current study is 86, or 10% of the faculty members who make up the entire population. By applying the (Krejcie & Morgan, 1970) formula, this sample size was determined. In this study, the faculty at private higher educational institutions serves as the analysis's unit of analysis. To derive an accurate inference from the results of the current study, the researcher employed the probability sampling technique, which enables the researcher to draw statistical conclusions about the entire population. One example of such a method is the proportionate systematic random sampling methodology, which was used in current research. First, it was decided that 10% of the faculty members from each private higher education institution would be included in the required sample size. The researcher has chosen a sampling fraction of 1/10, which is 10% of the population of each private higher education institute, considering the total number of samples proposed for the study as well as to ensure that the number of samples is proportionate to the total population of each private higher educational institutions This makes sure that the sample contains an equal and proportionate representation of all private higher education institutions in the target demographic. Each private higher education institution's sample proportion size is further separated into four age groups: 24–30, 31–36, 37–42. The survey was then randomly dispersed regardless of how many people live in each stratum. Two to four weeks were provided for the responders to complete the survey. The researcher followed up by recalling the respondents through their department heads for any questionnaires that were not returned within the allotted four weeks. However, it took over three months for all the respondents to finish the surveys.

A pilot test was carried out for this study before the revised and improved version of the research instrument was distributed to the respondents. For the pilot testing (Zhu et al., 2008). Suggested 25 to 100 responders, which might not also be selected statistically. The researcher asked two experts in educational leadership, management, and higher education to review and provide their opinions regarding the content of the survey questionnaire for this study through email to verify the authenticity of the content of the questionnaire. They were asked to review the questionnaire for any mistakes and unclear questions. After conducting a pilot test, the reliability of the instrument was determined to be 0.813 for the authentic leadership scale (AL) and 0.817 for the innovative work behavior scale (IWB). The sample respondents received the revised questionnaire afterward (Table 2).

Table 2. Hypotheses testing and beta coefficients.

Path	Antecedents	Beta	*p*	Conclusions
Ha1	Self-Awareness	.074	0.775	Rejected
Ha2	Internalized-Moral Perspective Balance Processing	.265	2.323	Supported
Ha3	Relational Transparency	.047	0.372	Rejected
Ha4	Authentic Leadership*Metaverse	.402	3.601	Supported
Ha5		.563	4.652	Supported

The table show that the coefficient results as indicate that the beta value of 0.074 which means that self-awareness has a weak effect on innovative work behaviour Compared to through independent variable. Moreover, that the beta value of 0.265 which means that internalized moral perspective has a moderate effect on innovative work behaviour Compared to through independent variable. Furthermore, balance processing coefficient that for indicate that the beta value 0.047 as indicating that weak effect on innovative work behaviour. Moreover, relational transparency coefficient indicate that the beta value 0.402 positively significant effects on innovative work behaviour. In summary internalized moral perspective and relational transparency have significant effect on innovative work behaviour while self-awareness and balance processing have no significant effect on innovative work behaviour. The predictor variable authentic leadership has an r square value 0.178 indicating that the independent variable includes the model explain approximately 17.8% of the variance in the independent variable innovative work behaviour. The relationship between authentic leadership and innovative work behavior was stronger when metaverse was high.

5 Discussions and Conclusions

The current study sought to investigate the impact of four distinct authentic leadership antecedents on faculty members' innovative work behaviors at private institutions in Pakistan's Lahore metropolis. Quantitative research methods were used to carry out the current study. Using SPSS version 25, the outcomes of the data collection were examined. According to the study's conclusions, "Internalized-Moral Perspective In contrast, "self-awareness" and "balance processing" have shown an insignificant influence on the innovative work behavior of faculty members of higher educational institutions. It can therefore be concluded that if faculty members of higher education institutions are satisfied with their work, they will remain with their higher education institutes for a long time. In private higher educational institutions, the faculty is more content if their leaders support their academic objectives, and they internalize a moral perspective and relational transparency. Most faculty members were happy with their leaders' relational honesty, then", "Relational Transparency ", whereas "self-awareness" and "Balance processing" have shown insignificant influence on the innovative work behavior of faculty members of higher education institutions, it can be concluded that if faculty of higher education institutes are satisfied with their work, they will stay intact with their higher education institutes for a long time, improving the quality of instruction. In private higher education institutions, the faculty is more content if their leaders support their academic objectives, and they internalize a moral perspective and relational transparency. Most faculty members were pleased with their leaders' interpersonal openness, which was followed by what they regarded to be an internalized moral stance. Ultimately, the Metaverse is an intriguing and revolutionary idea that has the potential to alter our relationships with digital surroundings, one another, and the physical world. A fully immersive, interconnected digital universe is getting closer to reality as technology develops. But solving many technical, ethical, and societal problems is necessary if the Metaverse is to reach its full potential. While virtual reality's future seems bright, its growth will benefit from thoughtful planning and good stewardship.

6 Limitations and Upcoming Studies

Only the city of Lahore was considered in the research. In the other cities of Pakistan, there are also many private higher educational institutions. To further validate the association between all variables, future study can be conducted in different geographical regions of Pakistan using the same model and one or more additional antecedents. A large sample size could not be researched due to logistical and time constraints as well. So, for more generalizability, further study can be done with a bigger sample size. To gain a deeper understanding of the connections between the research variables and innovative work behavior in higher education institutions, qualitative analysis can also be added.

References

Akram, T., Lei, S., Haider, M.J.: The impact of relational leadership on employee innovative work behavior in IT industry of China. Arab Econ. Bus. J. **11**(2), 153–161 (2016)

Alfaisal, R., Hashim, H., Azizan, U.H.: Metaverse system adoption in education: a systematic literature review. J. Comput. Educ., 1–45 (2022). https://doi.org/10.1007/s40692-022-00256-6

Alhmoudi, R.S., Singh, S.K., Caputo, F., Riso, T., Iandolo, F.: Corporate social responsibility and innovative work behavior: is it a matter of perceptions? Corp. Soc. Responsib. Environ. Manag. **29**(6), 2030–2037 (2022)

Allen-Ile, C., Mahembe, B., Balogun, T.V.: A confirmatory factor analytic study of an authentic leadership measure in Nigeria. SA J. Hum. Resour. Manag. **18**(1), 1–9 (2020)

Alnajjar, M., Hashim, J.: Innovative work behaviour induced by transformational leadership through altruism. Int. J. Work Innov. **2**(4), 257–283 (2020)

Azanza, G., Moriano, J.A., Molero, F.: Authentic leadership and organizational culture as drivers of employees' job satisfaction. Revista de Psicología del Trabajo y de las Organizaciones **29**(2), 45–50 (2013)

Bandura, C.T., Kavussanu, M., Ong, C.W.: Authentic leadership and task cohesion: The mediating role of trust and team sacrifice. Group Dyn. Theor. Res. Pract. **23**(3–4), 185 (2019)

Carter, S., Andersen, C., Turner, M., Gaunt, L.: "What about us?" Wellbeing of higher education librarians. J. Acad. Librariansh. **49**(1), 102619 (2023)

de Carvalho, L.P., Poleto, T., Ramos, C.C., de Assis Rodrigues, F., de Carvalho, V.D.H., Nepomuceno, T.C.C.: Predictors of digital competence of public university employees and the impact on innovative work behavior. Adm. Sci. **13**(5), 131 (2023)

Crawford, J.A., Dawkins, S., Martin, A., Lewis, G.: Putting the leader back into authentic leadership: reconceptualising and rethinking leaders. Aust. J. Manag. **45**(1), 114–133 (2020)

Crawford, J., Kelder, J.-A., Knox, M.W.: What does it take to be a social entrepreneur?: Authentic leaders and their effect on innovation. In: Leadership Styles, Innovation, and Social Entrepreneurship in the Era of Digitalization, pp. 282–310. IGI Global (2020)

De Jong, J., Den Hartog, D.: Measuring innovative work behaviour. Creativity Innov. Manage. **19**(1), 23–36 (2010)

Dincelli, E., Yayla, A.: Immersive virtual reality in the age of the Metaverse: a hybrid-narrative review based on the technology affordance perspective. J. Strateg. Inf. Syst. **31**(2), 101717 (2022)

Duarte, A.P., Ribeiro, N., Semedo, A.S., Gomes, D.R.: Authentic leadership and improved individual performance: affective commitment and individual creativity's sequential mediation. Front. Psychol. **12**, 675749 (2021)

Durrah, O., Kahwaji, A.: Chameleon leadership and innovative behavior in the health sector: the mediation role of job security. Empl. Responsibilities Rights J. **23**, 247–265 (2022)

Farid, T., et al.: The impact of authentic leadership on organizational citizenship behaviors: the mediating role of affective-and cognitive-based trust. Front. Psychol. **11**, 1975 (2020)

Faulks, B., Song, Y., Waiganjo, M., Obrenovic, B., Godinic, D.: Impact of empowering leadership, innovative work, and organizational learning readiness on sustainable economic performance: an empirical study of companies in Russia during the COVID-19 pandemic. Sustainability **13**(22), 12465 (2021)

Gai, T., Wu, J., Cao, M., Ji, F., Sun, Q., Zhou, M.: Trust chain driven bidirectional feedback mechanism in social network group decision making and its application in Metaverse virtual community. Exp. Syst. Appl. **228**, 120369 (2023)

Geltzer, J.: Fake news & film: how alternative facts influence the national discourse. Sw. L. Rev. **47**, 297 (2017)

Gokasar, I., Pamucar, D., Deveci, M., Gupta, B.B., Martinez, L., Castillo, O.: Metaverse integration alternatives of connected autonomous vehicles with self-powered sensors using fuzzy decision making model. Inf. Sci. **642**, 119192 (2023)

Grošelj, M., Černe, M., Penger, S., Grah, B.: Authentic and transformational leadership and innovative work behaviour: the moderating role of psychological empowerment. Eur. J. Innov. Manag. **24**(3), 677–706 (2021)

Hashim, K.L.: Enhancing innovative work behaviour of Malaysian public sector employees. Malays. J. Soc. Sci. Humanit. (MJSSH) **6**(2), 253–265 (2021)

Hudson-Smith, A., Batty, M.: Ubiquitous geographic information in the emergent Metaverse. Trans. GIS **26**(3), 1147–1157 (2022)

Isaksen, S.G.: Leadership's role in creative climate creation. In: Handbook of Research on Leadership and Creativity, pp. 131–158. Edward Elgar Publishing (2017)
Işık, C., et al.: The nexus between team culture, innovative work behaviour and tacit knowledge sharing: theory and evidence. Sustainability **13**(8), 4333 (2021)
Kaya, B., Karatepe, O.M.: Does servant leadership better explain work engagement, career satisfaction and adaptive performance than authentic leadership? Int. J. Contemp. Hosp. Manag. **32**(6), 2075–2095 (2020)
Kiersch, C., Peters, J.: Leadership from the inside out: student leadership development within authentic leadership and servant leadership frameworks. J. Leadersh. Educ. **16**(1), 148–168 (2017). https://doi.org/10.12806/V16/I1/T4
Kleynhans, D.J., Heyns, M.M., Stander, M.W.: Authentic leadership and follower trust in the leader: the effect of precariousness. SA J. Ind. Psychol. **47**(1), 1–10 (2021)
Knezović, E., Drkić, A.: Innovative work behavior in SMEs: the role of transformational leadership. Empl. Relat. Int. J. **43**(2), 398–415 (2021)
Krejcie, R.V., Morgan, D.W.: Determining sample size for research activities. Educ. Psychol. Measur. **30**(3), 607–610 (1970)
Latynina, A., Ryzhkova, O., Mikhailova, S.: Leadership in innovation: a case of the high-tech sector in the Krasnoyarsk Krai. In: 6th International Conference on Social, Economic, and Academic Leadership, ICSEAL-6-2019 (2020)
Maheswari, D., Ndruru, F.B.F., Rejeki, D.S., Moniaga, J.V., Jabar, B.A.: Systematic literature review on the usage of IoT in the Metaverse to support the education system. In: 2022 5th International Conference on Information and Communications Technology (ICOIACT), pp. 307–310. IEEE, August 2022
Mishra, P., Bhatnagar, J., Gupta, R., Wadsworth, S.M.: How work–family enrichment influence innovative work behavior: role of psychological capital and supervisory support. J. Manag. Organ. **25**(1), 58–80 (2019)
Mubarak, N., Khan, J., Yasmin, R., Osmadi, A.: The impact of a proactive personality on innovative work behavior: the role of work engagement and transformational leadership. Leadersh. Org. Dev. J. **42**(7), 989–1003 (2021)
Njoku, J.N., Nwakanma, C.I., Amaizu, G.C., Kim, D.S.: Prospects and challenges of Metaverse application in data-driven intelligent transportation systems. IET Intel. Transp. Syst. **17**(1), 1–21 (2023)
Novitasari, D., Siswanto, E., Purwanto, A., Fahmi, K.: Authentic leadership and innovation: what is the role of psychological capital? Int. J. Soc. Manage. Stud. **1**(1), 1–21 (2020)
Oh, J., Oh, S.: Authentic leadership and turnover intention: does organizational size matter? Leadersh. Org. Dev. J. **38**(7), 912–926 (2017)
Pamucar, D., et al.: Evaluation of metaverse integration alternatives of sharing economy in transportation using fuzzy Schweizer-Sklar based ordinal priority approach. Decis. Support. Syst. **171**, 113944 (2023)
Phuong, T.H., Takahashi, K.: The impact of authentic leadership on employee creativity in Vietnam: a mediating effect of psychological contract and moderating effects of subcultures. Asia Pac. Bus. Rev. **27**(1), 77–100 (2021)
Ribeiro, N., Gomes, D., Kurian, S.: Authentic leadership and performance: the mediating role of employees' affective commitment. Soc. Responsib. J. **14**(1), 213–225 (2018)
Saeed, S., Ali, R.: Relationship between authentic leadership and classroom management in public and private sector universities. J. Educ. Educ. Dev. **6**(1), 171–187 (2019)
Saxena, A., Prasad, A.: Exploring the influence of dimensions of workplace spirituality on innovative work behaviour: role of sense of God. Int. J. Ethics Syst. **39**(2), 183–212 (2023)
Sengupta, S., Bajaj, B., Singh, A., Sharma, S., Patel, P., Prikshat, V.: Innovative work behavior driving Indian startups go global–the role of authentic leadership and readiness for change. J. Organ. Chang. Manag. **36**(1), 162–179 (2023)

Sethibe, T., Steyn, R.: The impact of leadership styles and the components of leadership styles on innovative behaviour. Int. J. Innov. Manag. **21**(02), 1750015 (2017)

Teng, H.-Y., Yi, O.: How and when authentic leadership promotes prosocial service behaviors: a moderated mediation model. Int. J. Hosp. Manag. **104**, 103227 (2022)

Xi, N., Chen, J., Gama, F., Riar, M., Hamari, J.: The challenges of entering the metaverse: an experiment on the effect of extended reality on workload. Inf. Syst. Front. **25**(2), 659–680 (2023)

Xu, B.-D., Zhao, S.-K., Li, C.-R., Lin, C.-J.: Authentic leadership and employee creativity: testing the multilevel mediation model. Leadersh. Organ. Dev. J. **38**, 482–498 (2017)

Yamak, O.U., Eyupoglu, S.Z.: Authentic leadership and service innovative behavior: mediating role of proactive personality. SAGE Open **11**(1), 2158244021989629 (2021)

Ефимова, Г., Латышев, А.: Job satisfaction among employees of a higher educational institution. Вопросы образования/Educ. Stud. Moscow **1**, 72–108 (2023)

The Role of Virtual Reality Technology in Disclosing Future Information: Evidence from Iraqi Banks

Azhaar Al-Ali(✉) and Assmaa Mahdi Al-hashimi

Technical College of Management Kufa, Al-Furat Al-Awsat Technical University, Najaf, Iraq
azharally9@gmail.com, asmaa.mahdi@atu.edu.iq

Abstract. This study intends to evaluate the role of virtual reality (VR) technology to help predict future performance of the two Iraqi Bank based on the data for the years 2013–2022. The study is further interested to do any techniques exists in the context of VR those can improve Accounting Information System (AIS) of the studied banks. Researchers used an applied analytical method based on certain financial indicators to answer the propose research question. VR technology is dependent on the accuracy of information that can help investors to make investment decisions. There are always lots of information which a company has to relied on and information technology can help to manage these. Metaverse technology can be of great help. With the introduction of digital currency companies experience a change in their financial reports and digital technologies better manage financial reports. Metaverse technology will transform financial reporting and will create more interactive real time reports. On the basis of our findings, it is recommended that banks need to shift to the adoption of digital smart technologies to better manage financial reporting mechanisms.

Keywords: Future information · Virtual reality technology · Metaverse

1 Introduction

In recent years, researchers have started a debate about removing the boundary between the physical world and the digital world in an effort to improve the interaction between humans and machines (Manhal et al. 2023). This will facilitate a collaboration between the two in a flexible manner for improved performance. This can be achieved by taking advantage of Industry 4.0 technologies to build the world of the metaverse beyond the physical universe. Internet of things (IoT) is considered as an efficient technology of industry 4.0 that constitutes a metaverse (Smaili and Rancourt-Raymond 2022; Albahri et al. 2023). Similarly, artificial intelligence, blockchain, and cloud computing are also forms of metaverse where users experience a virtual world. These technologies help to perform activities those are not possible without the existence of the metaverse.

Accordingly, this study is intended to investigate the ways metaverse technologies can help to build an efficient AIS. This is believed that metaverse technologies can significantly improve AIS (Chukwuani 2022; AL-Fatlawey et al. 2021; Alhamdi et al. 2019),

M. Al-Emran et al. (Eds.): IMDC-IST 2024, LNNS 876, pp. 188–199, 2023.
https://doi.org/10.1007/978-3-031-51300-8_13

these benefits are realized in terms of in increasing the quality of inputs, operational processes, and the quality of outputs generated from a metaverse based AIS. There have been a number of studies to investigate the various aspects of metaverse. However, there is a dearth of studies on the use of metaverse technologies for AIS, especially in the banking sector (Marwa 2023). Hence, it is expected that this study will make a significant contribution to understand the role of metaverse technologies in the development of an efficient AIS in banking sector. In the light go forgoing discussion, this study poses following questions.

1. What are the technologies that can be included in the accounting information system that exists in the world of the metaverse?
2. What are the consequences of the accounting information system's reliance on these technologies in supporting the disclosure of future information?

The significance of this study is realized in terms of an addition to the already existing literature in general and in Arab countries in particular. The study will help to understand the role of metaverse technologies in an efficient AIS.

The study has got the following objectives. To identify the metaverse technologies that can be used for the AIS in banking sector of Iraq. To evaluate the consequences of the adoption of metaverse technologies for the AIS in banking sector of Iraq.

2 Theoretical Background

Industry 4.0 has transformed the industrial manufacturing process into smart ones. Japan has introduced the concept of fifth society. The fifth society as defined during the World Economic Forum in January 2019 is a vision to create a future society that help human through economic and technological progress to solves social problems (Mujiono 2021; Hesselbarth et al. 2023; Alnoor et al. 2022) with an integration of physical and virtual realities. The fifth society is a society of imagination where digital transformation combines creativity to solve problems and create new values (Muhsen et al. 2023; Al-Hchaimi et al. 2023; Chew et al. 2023). Hence, metaverse will completely transform human life into digital one and the same will be applied to the business models. Metaverse is characterized by its interoperability by considering it a unified framework or a pillar that connects many of the applications and services. This means that users will be able to interact simultaneously with multiple applications, in contrast to what is done when using desktop computers or mobile devices, i.e. it will be possible to navigate smoothly through thematic spaces (Pietro et al. 2021; Atiyah et al. 2023).

Virtual technology is defined as an interconnected network of social and networked environments in persistent multi-user platforms which enables the seamless embodiment of users in dynamic, real-time interactions with digital platforms (Mystakidis 2022; Jabbar et al. 2023; Zaidan et al. 2023). The degree of interactivity depends on the extent to which the system allows interactions and events by the user when interacting with the components of this system. It is the extent to which virtual reality tools and technologies are used that allow full immersion and direct interaction with virtual environments (Darwish 2017; Atiyah 2020). Participate is one of the important and distinctive features of virtual reality, especially those that are published on the international information network, and are designed according to a set of factors, including support for multiple users. Participatory features a group of individuals use the same virtual reality application at the same time (Abbas et al. 2023; Bozanic et al. 2023). It allows sharing a group of users of the virtual reality system at the same time so that each of them can interact alone or interact in the presence of others by performing certain tasks to finally achieve the educational goal of the virtual environment (Haitham 2018).

It is the user's sense of virtual reality environments and applications that he/she is completely surrounded by the components of this environment, and therefore he/she is inside this environment interacting as one of its components. The virtual reality environment is a dynamic environment that proceeds according to an integrated, successive systems, and the atmosphere of that environment is controlled by the element of self-control and dynamic movement (Pantelidis 2010; Eneizan et al. 2019). It is considered one of the most important characteristics of virtual environments, as the user here simulates the natural reality and real experience in an industrial-imaginary environment that does not exist in real reality, but rather they are tools and means that enable the user to simulate a specific environment in which he wants to learn as if he is inside this real environment (Saidin 2015; Wah et al. 2022).

Metaverse technology includes three components: the physical layer and consists of all the hardware to support the operational functions of the metaverse technology i.e., computing, communication and storage. A strong regular layer is of paramount importance to ensure scalability and ubiquitous access to the metaverse. The virtual layer provides a parallel living world through holograms for users through which they interact with each other or with other objects. The virtual layer must simulate and reflect real-world data and analytics in real-time using technologies such as digital twins' technology and the interaction layer, which acts as a bridge to connect users to the physical world with the virtual world. Users can upload their inputs into the physical world that are eventually translated into specific procedures in the virtual world (Grayscale 2021). All previous potential developments can be crowned with the stage of full coexistence with augmented reality, where devices and machines (robots, for example) are part of augmented reality. Augmented reality helps predict situations that are largely difficult to predict, as they require huge amounts of complex data. It involves a range of risks and requires very different tasks. For example, when crises or global developments occur, the required data can be obtained, which helps users to prepare instant reports and alert management instantly. Then they filter this data to reach the most important ones to identify fluctuations or events. Unexpected events surrounding the company are predicted by fetching data IoT technology that provides data or calls some indicators using this information. Augmented reality devices can be included in a library of procedures as potential solutions to the expected problems.

2.1 The Most Important Technologies of Virtual Reality

Virtual Currencies Technology: Intangible currencies are produced, stored, dealt with, and traded electronically in the global currency platforms through their estimated value in the market. These currency platforms operate outside the official monetary system and are available in digital form. Virtual currency units contain a reference number that does not repeat. Virtual currencies are considered one of the forms of digital authorized money because it appears in the digital form with the cryptographic feature or the use of ciphers and codes. It can be called encrypted currencies. Virtual currencies in financial transactions depend on the principle of peer-to-peer (P2P) and use a type of technology called topology i.e. topology of peer-to-peer networks (Sali 2021: 2002).

Blockchain: It is a system in which a record of transactions, especially those made in a cryptocurrency, is maintained across computers that are linked in a peer-to-peer network. Given that the series of blocks is managed by a protocol network to communicate between the contract and verify the health of the new blocks. There is no need for a central authority that controls all blocks. Something is published in the series of blocks for each person in the blockchain network, an equivalent to a copy of the blockchain (Marshall et al. 2021).

Cloud computing: It is a technology of computer services within the Internet. In a clearer and more comprehensive sense, we can say that the files and the basic part of the operating system and programs are located on the Internet. In its data centers, the customer called the subscriber and gets all or some of it according to the pay-as-you-go system that is often approved (Ctos 2017).

Artificial Intelligence Technology: It is a computational technique in artificial intelligence used as a new method to solve complex problems in the fields of machine learning, systems engineering, complex systems, market forecasting, optimization, nonlinear systems, financial analysis, humanities, and economics (Mirzaey 2017).

2.2 Hypotheses Development

The concept of future financial information, which is disclosed in the future financial statements, refers to information about the company's financial position, results of its operations, future cash expenditures, and changes in future equity, in addition to a summary of the accounting policies that have been used and important assumptions about the future. Information about the future that is prepared and presented in the form of complete financial statements or one or more elements of the financial statements that show forecasting of potential and future business results. The greater interest in accounting literature and practical importance of this information in influencing the decisions of its users when estimating the sustainability of corporate performance. (Abdul Ghaffar 2020). Future financial information is defined as any financial information about the future, and the information can be presented as complete financial statements or limited to one or more elements of the financial statements (Flood 2020). Meligy (2017) states that future financial information is that information that helps the users of the annual reports understand the company's future, estimate its future activities, and evaluate the management's ability to face the changes that arise in the future. Reported in annual reports, progress reports, and direct contact with financial analysts. Accounting disclosure is defined as a reflection of the financial information of a firm presented in the form of annual financial reports to help users' decisions (Mohamed et al. 2020). While electronic accounting disclosure is defined as an establishment of firms for one or more sites on the global information network with the aim of distributing and publishing financial and non-financial information and submitting annual and interim financial reports and performance reports using smart technological tools in order to make them available immediately and appropriately (Mendes-Da-Silva and Onusic 2012).

On the other hand, future information is defined as a type of information that includes management's expectations about future operating results, in addition to financial expectations such as subsequent year's profits, expected cash flows, and expected revenues. It also includes non-financial information such as strategies, risks, and uncertainties that can significantly affect actual results and cause deviations in the expected results (Aljiri and Hussainey 2007). Financial reports are considered as a tool for the communication process between the firms and users of financial reports. According to the conceptual framework for financial accounting issued by the Financial Accounting Standards Board the quality of financial reports refers to the quality of the information contained in these reports (Haouam 2020). For accounting information to be characterized by quality, it must have the qualitative characteristics of the accounting information. It must be appropriate and fairly express what it intends to. The usefulness of financial information improves if it is comparable, verifiable, timely and understandable. The basic qualitative characteristics of accounting information include relevance and faithful representation (Ferasso and Alnoor 2022). The information is considered appropriate if it can make a difference in the decision and of predictive and confirmatory value. Beneficial decisions are made by the different stakeholders." With technical progress and the advent of globalization, systems have changed and developed. The production of accurate, reliable, and correct information requires work on designing information systems that help control the huge amount of necessary information needed by economic units, thus facilitating the process of predicting future information which is in the interest of users. Information technology has greatly helped in that, as the data processing processes have become automated. Researchers expect that in light of the use of metaverse technology, major changes will occur in the disclosure process in financial reports as a result of the emergence of digital assets, as well as digital currencies and digital revenues. As a result of the use of metaverse technology will provide different forms of financial reporting and the ability for users to easily access and interact with it, and to search for data. This can create fundamental changes that will transform the current reporting practices into updated interactive reports real-time (Phomlaphatrachakom and Kalasindhu 2021; Atiyah et al. 2023).

An accounting information system is a system which is built to process financial information for better business decisions. It should be flexible to accommodate the changes in the business environment. It is believed that even with the adoption of metaverse technologies for AIS, the role of AIS will remain the same. A metaverse based AIS will help to process information more quickly to facilitate and improve business decisions. This study proposes the following hypotheses:

H1: Virtual reality technology has a positive impact on the accounting information systems in Iraqi banks.

3 Research Methodology and Data Analysis

Two banks were chosen for the purpose of analysis and comparison, namely the Bank of Baghdad and the Iraqi Investment Bank. The banks in the study sample applied virtual reality techniques, and the time period was chosen from 2013 to 2022. The study used a deductive approach to find answers to the proposed research questions. This approach has been used keeping in view the nature of the problem, hypotheses and the nature of information. The following are among the most important financial indicators through which profitability and risk can be measured are as follows (Click, 2023).

1. Net Profit Ratio: The ration can be calculated as in the equation below:

Net Profit Ratio = Net Profit/Sales * 100.

2. Net profit on capital: To measure the ability of working capital to achieve profitability, can be calculated as in the equation below:

Net profit on capital = net profit / working capital.

3. Return on Shareholders' Equity: A measure of the return on shareholders' investment and it can be calculated as in the equation below:

Return on Shareholders' Equity = Net Profit - Preferred Stock Dividends / Common Stock.

4. Dividend ratio: This ratio is of special importance to shareholders and potential investors. It is concerned with measuring the impact of the economic unit's performance on the prices of ordinary shares in the market. It can be calculated as in the equation below:

Dividend Ratio = Dividend / Earnings Realized After Taxes and Preferred Shares.

Table 1. Net Profit Percentage.

Year	Net profit after tax	Revenues	Net profit ratio
2013	8,875,000	17,249,000	51%
2014	13,916,000	29,946,000	46%
2015	15,440,000	28,870,000	53%
2016	2,170,000	7,111,000	31%
2017	1,487,000	5,806,000	26%
2018	544,920	3,701,000	15%
2019	767,000	3,608,000	21%
2020	1,452,650	4,865,000	30%
2021	−2,238,900	2,003,000	−112%
2022	5,692,000	7,767,000	73%

From Table 1, we notice that the banks profits decreased during the year (2021) by (112%), as the bank did not achieve high revenues during these years, due to not managing costs properly, which led to a decrease in its profits. On the other hand, the bank achieved the highest percentage of profits in the year (2022), as the percentage reached (73%). This indicates an increase in its activity and the expansion of its services, thus attracting investors, new customers, and maintaining its customer base.

It is evident from Table 2 that the ratio of net profit to working capital decreased during the year (2021), as the ratio of net profit to working capital reached (−0.90%). The reason for the decrease is due to fluctuation in work activity in the bank or the

Table 2. Indicator of the ratio of net profit to working capital

Year	Net profit after tax	Working capital	Ratio of net profit
2013	8875000	88903000	0.047
2014	13916000	275888000	0.0504
2015	15440000	268048000	0.0576
2016	2170000	276202000	0.0079
2017	1487000	263072000	0.0057
2018	544920	263596000	0.0021
2019	767000	242340000	0.0032
2020	1452650	249856000	0.0058
2021	−2238900	247922000	−0.009
2022	5692000	258575000	0.022

lack of services provided to customers, which consequently reduced the attractiveness of investors. We also noticed an increase in the net profit of the o working capital index during the year (2015) by (5.76%) in the fourth quarter. This means that the bank achieved profits during this year by increasing its banking activities and increasing its services provided to customers.

The value of R2 is (53%), which means that the independent variables included in the test (net profit after tax and revenues) explain (53%) the changes in the dependent variable, which is (net Profit ratio), meaning that (53%) of the changes in the net profit percentage are due to changes in the independent variables, and the rest of the changes are estimated at (47%), as they are due to other variables that were not included in the model.

It is clear from Table 3 in testing the parameters and the significant influence of the variables that revenue and net profit after tax are independent variables that are not statistically significant and that there are other variables that substantially affect the percentage of net profit but were not included in the model. This is evident through the value of (sig) being greater than (0.05). Continuing with the use of information technology, what other insights can be gained from the results of the table above? Can we get an answer to the question of what will happen? By building equations to predict the relationship of this question to the future, as shown below for each indicator.

$$\text{Profit (y)} = 0.274 - 2.615 + 7.098$$

From the above regression equation, it was found that revenue affects profitability in an inverse relationship, that is, every one unit increase in revenue leads to a decrease in profitability by (-2.615), while net profit after tax affects profitability in a direct relationship, that is, every one unit increase in net profit After tax, it leads to an increase in profitability by (7.098). It is clear from Table 3 in testing the parameters and the significant influence of the variables that net profit after tax and working capital are independent variables with statistical significance, and that they have a significant effect

Table 3. Analysis of transactions

Coefficients			B	T	Sig
Net profit ratio	Model (1)	(Constant)	0.274	3.674	0.008
		Revenues	−2.615	−1.190-	0.273
		Net profit after tax	7.098	1.717	0.130
Working capital ratio	Model (2)	(Constant)	2.043	18.343	0.000
		Net profit after tax	0.000	88.428	0.000
		working capital	0.000	−18.328	0.000
Return on equity ratio	Model (3)	(Constant)	4.398	8.813	0.000
		net profit after tax	0.000	128.497	0.000
		Ordinary shares	−.001-	−19.740-	0.000
		book value	−.668-	−2.161-	0.074
Interest coverage ratio	Model (4)	(Constant)	0.263	1.354	0.218
		net profit t before ax	0.000	−0.595	0.571
		Interest expenses	0.000	0.729	0.489

on the dependent variable (the ratio of net profit to working capital) and this is evident through the value of (sig) in being less than (0.05) and was equal to (0.00).

Ratio of net profit to working capital (y) = 2.043 + 3.755E−7−7.848E−9.

From the above regression equation, it was found that working capital affects profitability in an inverse relationship, that is, every one-unit increase in working capital leads to a decrease in profitability by (−7.848E−9), while net profit after tax affects profitability in a direct relationship, that is every one unit increase in net profit after tax leads to an increase in profitability by (3.755E). It is clear from Table 3 in testing the parameters and the significant effect of the variables that the net profit after tax and common shares are independent variables with a statistically significant significance, and that they have a significant impact on the dependent variable (the percentage of return on owners' equity), and this is evident through the value of (sig) as it was less than (0.05) and was equal to (0.00), while the book value variable was statistically insignificant because the value of (sig) was greater than (0.05) as it was equal to (0.074).

$$\text{Return on equity}(y) = 4.398 + 3.482E - 7 - 001 - 668$$

From the above regression equation, it was found that ordinary shares affect the return on equity in an inverse relationship, that is, every one-unit increase in working capital leads to a decrease in the return on equity by (−001), as well as the book value affects the return on equity in an inverse relationship That is, every one unit increase in the book value leads to a decrease in the return on equity by (−668), while the net profit after tax affects profitability in a direct relationship, that is, every one unit increase in the net profit after tax leads to an increase in profitability by (3.482E−7). Table 3 shown testing the parameters and the significant influence of the variables that the net profit tax before and interest expenses are independent variables that do not have a statistical significance, and do not have a significant effect on the dependent variable (interest coverage ratio). This is evident through the value of (sig) in being Greater than (0.05) for the two variables included in the model, which means that there are other variables that have a significant impact on the interest coverage ratio that were not included in the model.

$$\text{Interest coverage ratio (y)} = .263 - 1.796\text{E} + 9.178\text{E}{-8}$$

From the regression equation above, it turns out that net profit tax affects before interest coverage inversely, that is, every one-unit increase in net profit after tax leads to a decrease in the interest coverage ratio by (−1.796E), while interest expense affects interest coverage in a direct relationship. That is, every one-unit increase in interest expense leads to an increase in interest coverage by (9.178E−8).

4 Conclusions

The research aimed to study the effect of virtual reality technology on accounting information systems, especially in the process of future disclosure of information, the practice of its activities within this world, and the techniques on which the components of the accounting information system can be based in metaverse. The research, in its theoretical part, concluded the importance of metaverse for the economic units and selling digital and physical products, and innovating in producing new products by relying on many technologies. Virtual and augmented reality, which draws the line between the real world and the digital world. The study concluded that the metaverse is a digital version that expresses our tangible physical world, accessible through the Internet, and the use of some techniques that enable simulation of the work environment and building an immersive digital world and interactive experiences. The study also concluded the importance of basing the accounting information system on the technologies supported by metaverse. Its importance is realized in terms of improving quality of inputs, operational processes, and outputs of the accounting information system to obtain accounting data and information by linking the accounting information system to smart machines, processing and storing accounting data in the presence of a cloud-based database and data centers and recording transactions securely through the characteristics of transparency. The stability provided by the blockchain and thus builds future financial expectations that are transparent and appropriate in making economic decisions for users. The involvement and immersion of users in high-quality financial reports and the information they contain of a future nature. Our study recommends the need for economic units of various sizes and activities to adopt smart technology, and to issue accounting standards that regulate measurement and disclosure of aspects of smart technology, which provide guidance when preparing financial reports.

References

Abbas, S., et al.: Antecedents of trustworthiness of social commerce platforms: a case of rural communities using multi group SEM & MCDM methods. Electr. Commerce Res. Appl. 101322 (2023)

Abdel-Ghaffar, N.E.-S.M.: Using the data mining method to support the informational content of future financial statements and its impact on enhancing the efficiency of investment decisions in the Egyptian financial market - an applied study. J. Finan. Commer. Res. **21**(1) (2020)

Albahri, A.S., et al.: A systematic review of using deep learning technology in the steady-state visually evoked potential-based brain-computer interface applications: current trends and future trust methodology. Int. J. Telemed. Appl. (2023)

Al-Darwish (Ahmed), A.A.-A. (Rajaa): "Technological Innovations". Bin Abdullah Al-Darwish, Dar Al-Fikr Al-Arabi, M (2017)

AL-Fatlawey, M.H., Brias, A.K., Atiyah, A.G.: The role of strategic behavior in achievement the organizational excellence "Analytical research of the manager's views of Ur State Company at Thi-Qar Governorate". J. Administ. Econ. **10**(37) (2021)

Al-Gnbri, M.: Accounting and auditing in the metaverse world from a virtual reality perspective: future research. J. Metaverse **2**, 29–41 (2022)

Alhamdi, M., Alnoor, A., Eneizan, B., Abdulla, M., Abdulaali, A.R.: Determinants of the production system time (jit) on reduce waste: case study in a Salsal water company. Int. J. Acad. Res. Bus. Soc. Sci. **9**(7), 17–32 (2019)

Al-Hchaimi, A.A.J., Sulaiman, N.B., Mustafa, M.A.B., Mohtar, M.N.B., Hassan, S.L.B.M., Muhsen, Y.R.: A comprehensive evaluation approach for efficient countermeasure techniques against timing side-channel attack on MPSoC-based IoT using multi-criteria decision-making methods. Egyptian Inf. J. **24**(2), 351–364 (2023)

Alnoor, A., et al.: How positive and negative electronic word of mouth (eWOM) affects customers' intention to use social commerce? A dual-stage multi group-SEM and ANN analysis. Int. J. Hum.–Comput. Interact. 1–30 (2022)

Al-Naimi, A.H.M.: Using neural networks in identifying audit risks / a case study in the Federal Financial Supervision Bureau. Higher diploma research in auditing and auditing in accounting. University of Mosul, College of Administration and Economics, Department of Accounting (2019)

Ashour a.n Mostafa. Review of Data Mining Concept and its Techniques (2018). www.researchgate.net/publication/323108632

Atiyah, A.G.: Impact of knowledge workers characteristics in promoting organizational creativity: an applied study in a sample of Smart organizations. PalArch's J. Archaeol. Egypt/Egyptol. **17**(6), 16626–16637 (2020)

Atiyah, A.G.: Power Distance and Strategic Decision Implementation: Exploring the Moderative Influence of Organizational Context

Atiyah, A.G.: Strategic Network and Psychological Contract Breach: The Mediating Effect of Role Ambiguity

Boukhazna, B.S., Ayman, I.B.Q.A.M.B.: "The role of electronic disclosure in activating corporate governance and its impact on improving the quality of financial reports," a supplementary memorandum that falls within the requirements for obtaining a master's degree in financial and accounting sciences, specializing in accounting and auditing (2021)

Bozanic, D., Tešić, D., Puška, A., Štilić, A., Muhsen, Y.R.: Ranking challenges, risks and threats using Fuzzy Inference System. Decision Making Appl. Manage. Eng. **6**(2), 933–947 (2023)

Bozzolan, S., et al.: Forward-looking disclosures, financial verifiability and analysts' forecasts: a study of cross-listed European firms. European Accounting Review **18**(3), 435–473 (2009)

Chew, X., Khaw, K.W., Alnoor, A., Ferasso, M., Al Halbusi, H., Muhsen, Y.R.: Circular economy of medical waste: novel intelligent medical waste management framework based on extension linear Diophantine fuzzy FDOSM and neural network approach. Environ. Sci. Pollut. Res. 1–27 (2023)

Chukwuani, V.: Virtual reality and augmented reality: its impact in the field of accounting. Contemporary Journal of Management, 35–40, ISSN 2766–1431, Published by AIR JOURNALS (2022). https://airjournal.org/cjm

Abdel-Masih, S.S.F.:"Investing in Virtual Currencies" The Legal Journal (a journal specialized in legal studies and research) A peer-reviewed scientific journal (ISSN: 2537-0758) (2021)

Eneizan, B.M., AL-kharabsheh, K.A., AL-Abrrow, H., Alnoor, A.: An Investigation into the Relationship between Emotional Labor and Customer Satisfaction. Central Eur. Manage. J. **27**(4), 23–47 (2019)

Ferasso, M., Alnoor, A.: Artificial neural network and structural equation modeling in the future. In: Artificial Neural Networks and Structural Equation Modeling: Marketing and Consumer Research Applications, pp. 327–341. Springer, Singapore (2022)

Flood, J.M.: Practitioner's Guide to GAAS 2020 Covering All SASS SSAES SSARSS and Interpretations. Wiley, Canada (2020)

Grayscale 2021. THE METAVERSE Web 3.0 Virtual Cloud Economies.: https://grayscale.com/wp-content/uploads/2021/11/GrayscaleMetaverse_Report Nov2021.pdf

Haouam, D.: IT governance impact on financial reporting quality using COBIT framework. Global J. Comput. Sci. Theory Res. **10**(1), 1–10 (2020)

Hassan (Haitham): Virtual World Technology and Augmented Reality in Education. Al-Khamayel: Arab Academic Center (2018)

Hesselbarth, I., Alnoor, A., Tiberius, V.: Behavioral strategy: A systematic literature review and research framework. Management Decision (2023)

Hind, M.H.Q.: Using neural networks - artificial intelligence - in predicting future economic growth in Egypt. SUST Journal of Future Studies (2016) ISSN (text): 1858 - 7003 Vol. (02) (2016)

https://www.emerald.com/insight/0265-2323.htm

Jabbar, A.K., Almayyahi, A.R.A., Ali, I.M., Alnoor, A.: Mitigating uncertainty in the boardroom: Analysis to financial reporting for financial risk COVID-19. J. Asian Fin. Econ. Bus. (JAFEB) **7**(12), 233–243 (2020)

Alijifri, K., Hussainey, K.: The determinants of forward-looking information in annual reports of UAE companies. Manage. Audit. J. **22**(9), 881–894 (2007)

Ashour, K., Jazia, H.: Ways to take advantage of cloud computing to protect electronic banking operations. Chlef University, Algeria, North African Economics J. (12) (2017)

Keung, E.C.: Do supplementary sales forecasts increase the credibility of financial analysts earnings forecasts? Account. Rev. **85**(6), 2047–2074 (2010)

Mohamed, K.A.S., Al-Waeli, A.J., Rasheed, H.S.: The role of disclosure of future financial information in maximizing the value of the company in the Iraqi industrial companies. International Journal of Management (LIM) **11**(11)

Mackenzie, K., Buckby, S., Irvine, H.: Business research in virtual worlds: possibilities and practicalities. Acc. Aud. Account. J. **26**(3), 352–373 (2013). https://doi.org/10.1108/09513571311311856

Manhal, M., Al-khalidi, A., Hamad, Z.: Strategic network: managerial myopia point of view. Manage. Sci. Lett. **13**(3), 211–218 (2023)

Romney, M.B., Steinbart, P.J., Summers, S.L., Wood, D.A.: Accounting Information Systems-Global Edition (2021)

Maryam, A.-R., Al-Qashawi, A.R.: An analytical study to evaluate the role of artificial intelligence techniques in improving the electronic accounting disclosure process. J. Adm. Fin. Quant. Res. Part 2, 2 (2022)

Muhsen, Y.R., Husin, N.A., Zolkepli, M.B., Manshor, N., Al-Hchaimi, A.A.J.: Evaluation of the routing algorithms for NoC-based MPSoC: a fuzzy multi-criteria decision-making approach. IEEE Access **11**, 102806–102827 (2023)

Mujiono, M.N.: The shifting role of accountants in the era of digital disruption. In: Educational Research and Business Conference (ERABC), 1–14 (2021). Narin, N. (2021). A Content Analysis of the Metaverse. Journal of Metaverse 1 (1). 17–24. www.secondlife.com

Mystakidis, S.: Metaverse. Encyclopedia **2**(1), 486–497 (2022)
Bamodu, O., Ye, X.: Virtual reality and virtual reality system components In: Proceedings of the 2nd International Conference on -- Systems Fugiting and Modeling (CSFM-13) Faculty of Computing, Engineering and Technology, Staffordshire University, United Kingdom (2013)
Pantelidis, V.: "Reasons to use virtual reality in -3 education and training courses and a model to determine when to use virtual reality. Themes Sci. Technol. Educ. 59–70 (2010)
Phornlaphatrachakorn, K., Na Kalasindhu, K.: Digital accounting, financial reporting quality and digital transformation: Evidence from Thai listed firms. J. Asian Fin. Econ. Bus. **8**(8), 409–419 (2021)
Pietro, R., Cresci, S.: Metaverse: Security and Privacy Issues. In: Proceedings - 2021 3rd IEEE International Conference on Trust, Privacy and Security in Intelligent Systems and Applications, TPS-ISA 2021, pp. 281–288 (2021). https://doi.org/10.1109/TPSISA52974.2021.00032
Republic Realm: The 2021 Metaverse Real Estate Report (2021). https://www.republicrealm.com/post/the-2021-metaverse-real-estate-report
Saidin. N., Abd Halim. N., Yahaya. N.: "A -4 review of research on augmented reality in education: advantages and applications. Int. Educ. Stud. **8**(13) (2015)
Saverio: Quality versus quantity: the case of forward-looking disclosure, Department of Economic Sciences-University of Padova Via del Santo 33 (2008)
Smaili, N., de Rancourt-Raymond, A.: Metaverse: welcome to the new fraud marketplace. J. Finan. Crime 1–13 (2022). https://doi.org/10.1108/JFC-06- 2022-0124
Wah, K.K., Omar, A.Z., Alnoor, A., Alshamkhani, M.T.: Technology application in tourism in Asia: comprehensive science mapping analysis. In Technology Application in Tourism in Asia: Innovations, Theories and Practices, pp. 53–66. Springer, Singapore (2022)
Mendes-Da-Silva, W., Onusic, L.M.: Corporate e-disclosure determinants: evidence from the Brazilian market. In: International Journal of Disclosure and Governance advance online publication, 23 August (2012)
Zaidan, A.A., et al.: Review of artificial neural networks-contribution methods integrated with structural equation modeling and multi-criteria decision analysis for selection customization. Eng. Appl. Artif. ,. **124**, 106643 (2023)

The Effect of Marketing Mood Management in Enhancing Sustainability: Evidence from Virtual Marketing Platforms

Mortada Mohsen Taher Al-Taie(✉), Ilham Nazem Abdel-Hadi, and Hossam Hussein Shiaa

Faculty of Administration and Economics, University of Karbala, Karbala, Iraq
altaeemortadha@gmail.com, {ilham.nadhom, hussam.h}@uokerbala.edu.iq

Abstract. The main goal of the current research is to test how marketing mood management can affect enhancing sustainability. In order to answer the research questions and achieve its objectives, virtual marketing platforms such as Facebook and Roblox users were adopted as a field for implementing the research. The research population consisting of users of virtual marketing platforms numbering (135) as a sample responding to the research. After collecting data and information through a questionnaire from the research sample, it was analyzed using (SPSS V. 25) and (SmartPLS 3) programs. A set of conclusions were reached, the most important of which was that there is a significant effect of marketing mood management in enhancing sustainability between users of virtual reality platforms.

Keywords: Marketing Mood Management · Sustainability · Virtual Reality

1 Introduction

The concept of sustainability has become critical in the modern world, as it reflects the global shift towards more sustainable practices and an orientation towards sustainable development (Demastus and Landrum 2023). Sustainability embodies organizations' commitment to achieving a balance between continuous improvement and social and environmental responsibility (Bilderback 2023). Sustainability includes a comprehensive approach to managing resources effectively, achieving efficiency in operations, and enhancing sustainable communication with all stakeholders (Choi et al. 2023; Hamid et al. 2021). These concepts reflect a commitment to sustainable participation in the economy, maintaining environmental balance, and promoting social justice (Dwivedi et al. 2023; Al-Abrrow et al. 2021). Organizations that adopt the principles of sustainability work to develop long-term strategies that respond to contemporary challenges and meet the aspirations of current and future generations. It requires a shift in work culture and management methods to integrate economic, social, environmental, cognitive and competitive dimensions into every aspect of organizational work (Roos et al.

M. Al-Emran et al. (Eds.): IMDC-IST 2024, LNNS 876, pp. 200–210, 2023.
https://doi.org/10.1007/978-3-031-51300-8_14

2023; Khaw et al. 2022). With increasing awareness of the importance of sustainability, organizations committed to its principles are able to build a strong reputation and achieve competitive advantage. Sustainability represents a challenge that requires organizations to be catalysts for positive change and sustainable development in the communities they serve (Gomes et al. 2023; Eneizan et al. 2019; Alsalem et al. 2022).

In order to achieve sustainability for business organizations at the present time, the most important pillar that they must pay attention to is the customer (Albahri et al. 2021). This is because the customer is the mainstay of the organization's survival and continuation of its work (Bestman et al. 2022; Fadhil et al. 2021). Thus, a large group of concepts and tools have emerged through which the relationship with customers dealing with the organization can be maintained, including what is known today as "marketing mood management". Marketing mood management is an aspect of modern strategies that is becoming more important in the business and marketing arena. This management relates to how to form and manage the impressions and feelings of customers and the target audience. This approach requires a deep understanding of customers' preferences and expectations and the ability to generate positive responses and effective communication (Yin and Chang 2022; Alnoor et al. 2022). Marketing mood management includes a set of strategies and techniques to improve customers' experience and strengthen the connections between them and the organization's products and brand (AL-Fatlawey et al. 2021). By creating a positive mood and promoting positive emotions, organizations can enhance customer loyalty and increase their value in the long term (Bengtsson and Johansson 2022). In a world full of competition, marketing mood management enhances customer interaction with the brand and contributes to enhancing social and digital interactions (Muhsen et al. 2023; Al-Hchaimi et al. 2023; Chew et al. 2023). By using techniques such as emotional marketing and powerful stories, organizations can achieve profound positive impacts on customer decisions and behaviors (Abbas et al. 2023; Bozanic et al. 2023). Directing efforts towards enriching the customer experience with positive emotions and good feelings, organizations develop strong and sustainable relationships with their customers and achieve superiority in the competitive arena (Dakir et al. 2022; Gatea and Marina 2016). The current research contributes to bridging the knowledge gap between the studied variables, as the relationships between these variables have not been adequately studied in previous studies, especially the two variables marketing mood management and sustainability in virtual reality platforms (Manhal et al. 2023). Hence, the current research provides a new vision and comprehensive analysis of the mutual influences between these variables.

2 Literature Review and Development of Hypotheses

2.1 The Concept of Marketing Mood Management

Mood is an emotional state that individuals subjectively perceive and ranges from positive to negative. A positive mood consists of feelings of increased happiness, and conversely, a negative mood consists of feelings of increased dissatisfaction (Atiyah and Zaidan 2022). This results when customers purchase pleasant or unpleasant products and their impact on their mood changes from positive to negative or vice versa. Mood can be manipulated in a variety of promotional and advertising methods, which has been proven in studies and research conducted by many writers, researchers and practitioners in this field. Individuals can be subjected to imaginary reactions by marketers by providing them with information as a means of influencing a positive or negative mood. That is, other individuals can manage the mood that forms in individuals, and this management may be negative, which affects the customer's purchasing behavior (Noland 2021, Atiyah 2023). The mood of customers can be managed and manipulated by marketers (whether employees on the front lines or employees on websites), through some tactics that are being used. This process is known as "marketing mood management." In order to successfully manage mood, many researchers have focused on classifying customers according to their good or bad mood in order to manage them better. Customers who are in a good mood, are likely to be happy easily. Because these types of customers are more sympathetic than people in a bad mood. While happiness, cheerfulness, serenity, and warmth show positive mood, unhappiness, anxiety, guilt, and depression are associated with negative mood (Sirakaya et al. 2004).

In the same context, (Kocabulut and Albayrak 2019) believe that although marketing managers cannot directly control the mood of customers, by offering some special gifts or amazing experiences, managers and employees may try to change the negative mood of customers into a negative mood. Positive. In addition, managers may prioritize hiring people who have good communication skills. For current employees, training programs can be organized aimed at strengthening relations between employees and customers. Improving the quality of service and people-based communications may change the bad and negative mood of customers into a positive one. Given the novelty of the topic of marketing mood management, and the scarcity of studies dealing with it (as far as the researcher knows), there have not been many studies addressing the concept of this topic with some clarity and detail. Therefore, some definitions of marketing mood management will be included according to the opinions of some researchers, as follows. Marketing mood management was defined by (Robinson & Knobloch-Westerwick 2016; Atiyah 2020) as the process of managing individuals' moods that stimulate their behaviors that lead to changes in their mood and thus change their intention and acceptance of products and services, which leads to changes in behavior. While (Davydov et al. 2021) defined it as the process in which marketers use behavioral and cognitive methods to manage the negative moods of individual consumers with the aim of swaying their consumption behavior and enhancing their purchase intention.

2.2 Marketing Mood Management and Virtual Reality Platforms

Marketing mood management is a concept that involves using various marketing strategies and techniques to influence and manage the emotional state or mood of consumers. This is done to create a positive and engaging experience that resonates with the target audience. Virtual Reality (VR) platforms have emerged as a powerful tool in achieving mood management in marketing (Yung and Khoo-Lattimore 2019). VR platforms are immersive digital environments that use technology to simulate a three-dimensional, computer-generated world, allowing users to interact with and experience this virtual world as if it were real. VR platforms encompass hardware and software components designed to create a convincing and immersive experience for users (Hollebeek et al. 2020). VR platforms offer a unique opportunity to connect with consumers on a deep emotional level, leveraging the immersive and interactive nature of virtual reality (Alcañiz et al. 2019). By utilizing VR in marketing, brands can achieve the following:

a. Empathy and Connection: VR can transport consumers to different environments and situations, allowing them to experience the brand's message firsthand. This can foster empathy and a sense of connection between the consumer and the brand, leading to a positive emotional response (Brydon et al. 2021).
b. Memory and Recall: Immersive experiences created through VR are more likely to be remembered and recalled. By associating positive emotions with their brand, marketers can create lasting impressions in consumers' minds, increasing brand recall and loyalty (Boyd and Koles 2019).
c. Consumer Engagement: VR experiences are highly engaging, capturing the full attention of users. This increased engagement can boost the effectiveness of marketing campaigns, as consumers are more likely to absorb and respond positively to the message (Violante et al. 2019).
d. Influence on Purchase Decisions: When consumers are in a positive mood, they are more receptive to marketing messages and more likely to make purchase decisions. VR can be used to create enjoyable and memorable interactions that lead to conversions (Kang et al., 2020).
e. Brand Differentiation: Utilizing VR sets a brand apart from competitors. It demonstrates innovation and a commitment to providing unique and enjoyable experiences for customers, which can enhance brand perception and loyalty (Moriuchi et al., 2021).
f. Global Reach: VR transcends geographical boundaries, allowing brands to connect with a global audience. This can lead to an expanded customer base and increased market share (Farshid et al., 2018).
g. Measurable Results: VR marketing campaigns can provide valuable data on user interactions and emotional responses. Marketers can use this data to fine-tune their strategies for better mood management and improved campaign performance (Cook et al. 2019).
h. Storytelling Opportunities: VR enables brands to tell immersive and impactful stories. This storytelling can evoke a wide range of emotions, from excitement and joy to empathy and inspiration, all of which can positively influence consumer moods (Barreda-Ángeles et al. 2021).

Marketing mood management included four main dimensions namely: Operant conditioning, the process that marketing personnel use to shape customer behavior in order to increase sales or achieve other desired results. Reducing negative marketing stimuli, the process through which front-line employees eliminate negative incentives that reduce consumer-purchasing behavior and enhance their acceptance of the product or service provided by the organization. Reducing marketing depression, the process through which the depression that the customer faces regarding the organization's products or services or its marketing activities is eliminated or reduced. Emotional excitement, the process by which front-line employees attempt to influence customers' emotions in order to attract them toward purchasing a product or service by arousing positive emotion. Thus, we assume that:

H1: There is a relationship between marketing mood management and sustainability of users of virtual marketing platforms.

2.3 The Concept of Sustainability

Sustainability has become of great importance and a high priority in today's business environment for practitioners and researchers. Successful implementation of sustainability practices has become the cornerstone of organizational survival, success, superior performance and competitive advantages. To be sustainable, organizations must define business models on how to modify organizational processes. Given that business organizations both contribute to and are affected by these sustainability challenges, management scholars are calling for increased research interest in how organizations address these issues (Magd and Karyamsetty 2021). While Smilevski (2017) pointed out that sustainability is one of the necessary and always desirable situations and situations in the life of any organization.

Rao and Jha (2020) explained that sustainability requires committed employees and is therefore required to adopt new innovative ways of doing things and create feelings such as fun, excitement, immersion, and engagement in real-world situations. This will help develop creativity and build relationships within organizations. On the other hand, it may help increase employees' work engagement, loyalty and commitment, along with problem-solving abilities and attract, motivate, train, reward, engage and retain employees to improve sustainability. Regarding the concept of sustainability, there have been many definitions of this concept according to the opinions of many writers and researchers. Sustainability has been defined by (Afrazeh et al. 2010) as the ability of an organization to maintain or develop its performance in the long term and is the result of stakeholder satisfaction during that time. While (Oliari et al., 2021) defined it as organizational activities that proactively seek to contribute to achieving a sustainable balance in the economic, environmental and social aspects in the short, medium and long term.

Sustainability consists of five dimensions, which economic sustainability, the way organizations should make available resources in the long term to maximize profits and ensure adequate revenues for stakeholders (Amornkitvikai et al. 2022). Social sustainability, developing a sustainable environment that improves living standards and social well-being by paying attention to the needs of all stakeholders (Lahri et al. 2021). Environmental sustainability, meeting the resource and service needs of current and future generations without compromising the health of the ecosystems they provide, and more specifically, as a condition for the balance, resilience and interconnectedness that allows human society to be satisfied (Özel, 2022). Knowledge sustainability, the process in which an organization maintains and develops data, information, and knowledge by focusing on useful and important information and reducing unnecessary information (Vágner and Bencsik, 2022). Competitive sustainability, the mechanism for achieving sustainable development through harmonizing local and international environmental standards through the use of competitive forces that reward the most efficient economic actors (Lopez & Brands, 2010).

3 Method

The research sample consisted of (135) users of virtual shopping platforms in the city of Baghdad. The current study used a number of previous literatures to measure the study variables. A five-point Likert scale was used. Moreover, the scale of Rizomyliotis et al. was adopted. (2018) to measure marketing mood. While Abumalloh et al. (2023) to measure sustainability in virtual shopping networks. To address the common problem of bias, Harman's one factor test was used. The results confirmed that the rate of variance extracted was less than 50%, which confirms that there is no concern about bias in the data. The sample characteristics of the study were that the percentage of females was greater than that of males. The percentage of females was 61% and the percentage of males was 39%. The group from 35 to 45 obtained the largest percentage, reaching 48%. The average customer's experience in purchasing via virtual networks was more than ten years.

4 Results

The results of the descriptive analysis of the dimensions of the independent variable marketing mood management, shown in Table 1 revealed that there are varying levels of prevalence of these dimensions in virtual shopping platforms. It was also shown that the dimensions carry different ordinal importance that are close to each other. From the results, it is clear that the dimension of reducing negative marketing stimuli is the most widespread, as it obtained an arithmetic mean of (4.16) and a standard deviation of (.674). While the emotional excitement dimension was the least widespread among the dimensions of the variable, its arithmetic mean was (4.01), and its standard deviation was (.844).

Table 1. Descriptive analysis of marketing mood management.

No	Dimensions of Marketing Mood Management	Mean	Std	R. Imp	Sequence
1	Operant Conditioning	4.08	.814	82%	2
2	Reducing negative marketing stimuli	4.16	.674	83%	1
3	Reducing marketing depression	4.06	.728	81%	3
4	Emotional excitement	4.01	.844	80%	4
Total of Marketing Mood Management		4.08	.397	82%	First

The results of the descriptive analysis of the dimensions of the dependent variable (sustainability), shown in Table 2 revealed that there are varying levels of prevalence of these dimensions in virtual shopping platforms. It was also shown that the dimensions carry different ordinal importance that are close to each other. From the results, it is clear that the dimension of cognitive sustainability is the most widespread, as it obtained an arithmetic mean of (4.24) and a standard deviation of (.569). While the environmental sustainability dimension was the least widespread among the dimensions of the variable, its arithmetic mean was (3.83), and its standard deviation was (.871).

Table 2. Descriptive analysis of organizational sustainability.

No.	Dimensions of Organizational Sustainability	Mean	Std	R. Imp	Sequence
1	Economic sustainability	3.86	.894	77%	4
2	Social sustainability	4.05	.766	81%	3
3	Environmental sustainability	3.83	.871	76%	5
4	Knowledge sustainability	4.24	.569	85%	1
	Competitive sustainability	4.21	.522	84%	2
Total of Organizational Sustainability		4.07	.607	81%	Second

In order to test this hypothesis, we used SmartPLS3 program, and built a structural model in order to show the relationship path of the study variables represented by (marketing mood management) as an independent variable, and (sustainability) as a dependent variable.

Table 3. Results of the analysis of the main research hypothesis.

Path	β	R^2	T. Value	Sig	Result
Marketing Mood Management → Organizational Sustainability	934	873	9.578	000	Accept

By looking at Fig. 1 of the structural equation model related to testing the main hypothesis of the research, and Table 3 of the results of the path analysis, the results show the possibility of accepting the main hypothesis of the research, which states: "There is a significant effect of marketing mood management on enhancing organizational sustainability". Where the value of the effect coefficient (β) reached (.934), and this effect is significant because the level of significance was (.000), which is less than the level of significance assumed by the researcher (0.05). In addition, the marketing mood management variable explains (87.3%) of all changes that can occur in organizational sustainability variable, a value indicated by the Explanation Coefficient (R^2). To make sure that the hypothesis was accepted or not, the (T) value test was implemented, and after analyzing it, the (T) value reached (9.578), which is greater than its tabular value of (2.00). Based on these results, the main hypothesis of the research is accepted.

5 Discussion and Conclusion

It was found that there is a relatively high availability of the dimensions of marketing mood management in virtual shopping platforms users. The arithmetic means for this variable, which was extracted based on the (SPSS V.25) program, was (4.08), which indicates clear superiority in applying marketing mood management strategies, as this number is considered an indication of the serious efforts made by these centers to interact with the needs and expectations. Customers effectively. There is a relatively high degree of effort towards enhancing sustainability on the part of virtual shopping platforms users. The arithmetic means of this variable, which was extracted based on the (SPSS V.25) program, was (4.07), which indicates the willingness of these platforms to adopt sustainable practices and strategies, which can contribute to enhancing their competitiveness and achieving their goals in the long term, including providing better customer services and maintaining environmental, social and economic balance. In addition, there is a significant effect of marketing mood management on enhancing sustainability in virtual shopping platforms users. This is based on the value of the effect factor (β) extracted using the (SmartPLS3) program, which is (.934), which is positive and significant, as the achieved significance level reached (000), which is less than the significance level assumed by the researcher, that amount is (0.05).

In conclusion, this research highlighted the impact that marketing mood management can play in enhancing sustainability in the contemporary business world. Therefore, the research aimed to test the extent of the impact of marketing mood management in enhancing sustainability in virtual shopping platforms users. The results of this research show that marketing mood management constitutes a vital element in the contemporary business context. It contributes significantly to enhancing sustainability by directing attention towards meeting the needs and expectations of customers effectively and motivating them to interact positively with the products and services provided. Marketing operations based on mood management work to enhance customer satisfaction and thus increase loyalty and revenues. Accordingly, marketing mood management can have a pivotal role in business strategies and organizational planning for virtual shopping platforms users. Motivating marketers to adopt this approach and develop marketing mood management skills can have a positive impact on the platforms' overall performance and their ability to compete in an increasingly competitive market. The researchers hope that this research will be an indication to leaders and decision-makers in commercial institutions of the importance of considering the role of marketing mood management as a powerful strategic tool to enhance sustainability and sustainable growth and achieve the goals of success in the business world today and in the future.

References

Abbas, S., et al.: Antecedents of Trustworthiness of Social Commerce Platforms: A Case of Rural Communities Using Multi Group SEM & MCDM Methods. Electron. Commerce Res. Appl. 101322 (2023)

Abumalloh, R.A., et al.: The adoption of metaverse in the retail industry and its impact on sustainable competitive advantage: moderating impact of sustainability commitment. Ann. Oper. Res. 1–42 (2023)

Afrazeh, A., Mohammadnabi, S., Mohammadnabi, S.: Organizational sustainability assessment and improvement model with knowledge management approach. Manage. Stud. Dev. Evol. **20**(61), 37–63 (2010)

Al-Abrrow, H., Fayez, A.S., Abdullah, H., Khaw, K.W., Alnoor, A., Rexhepi, G.: Effect of open-mindedness and humble behavior on innovation: mediator role of learning. Int. J. Emer. Markets (2021)

Albahri, A.S., et al.: Based on the multi-assessment model: towards a new context of combining the artificial neural network and structural equation modelling: a review. Chaos Solit. Fractals **153**, 111445 (2021)

Alcañiz, M., Bigné, E., Guixeres, J.: Virtual reality in marketing: a framework, review, and research agenda. Front. Psychol. **10**, 1530 (2019)

AL-Fatlawey, M.H., Brias, A.K., Atiyah, A.G.: The role of strategic behavior in achievement the organizational excellence "Analytical research of the manager's views of Ur State Company at Thi-Qar Governorate". J. Adm. Econ. **10**(37) (2021)

Al-Hchaimi, A.A.J., Sulaiman, N.B., Mustafa, M.A.B., Mohtar, M.N.B., Hassan, S.L.B.M., Muhsen, Y.R.: A comprehensive evaluation approach for efficient countermeasure techniques against timing side-channel attack on MPSoC-based IoT using multi-criteria decision-making methods. Egyptian Inf. J. **24**(2), 351–364 (2023)

Alnoor, A., et al.: How positive and negative electronic word of mouth (eWOM) affects customers' intention to use social commerce? A dual-stage multi group-SEM and ANN analysis. Int. J. Hum.–Comput. Interact. 1–30 (2022)

Alsalem, M.A., et al.: Rise of multiattribute decision-making in combating COVID-19: a systematic review of the state-of-the-art literature. Int. J. Intell. Syst. **37**(6), 3514–3624 (2022)

Amornkitvikai, Y., Tham, S.Y., Harvie, C., Buachoom, W.W.: Barriers and factors affecting the e-commerce sustainability of Thai micro-, small-and medium-sized enterprises (MSMEs). Sustainability **14**(14), 8476 (2022)

Atiyah, A.G.: The effect of the dimensions of strategic change on organizational performance level. PalArch's J. Archaeol. Egypt/Egyptol. **17**(8), 1269–1282 (2020)

Atiyah, A.G.: Strategic network and psychological contract breach: the mediating effect of role ambiguity

Atiyah, A.G., Zaidan, R.A.: Barriers to using social commerce. In: Artificial Neural Networks and Structural Equation Modeling: Marketing and Consumer Research Applications, pp. 115–130. Springer, Singapore (2022). https://doi.org/10.1007/978-981-19-6509-8_7

Barreda-Ángeles, M., Aleix-Guillaume, S., Pereda-Baños, A.: Virtual reality storytelling as a double-edged sword: immersive presentation of nonfiction 360-video is associated with impaired cognitive information processing. Commun. Monogr. **88**(2), 154–173 (2021)

Bengtsson, S., Johansson, S.: The meanings of social media use in everyday life: Filling empty slots, everyday transformations, and mood management. Social Media+ Society **8**(4), 20563051221130292 (2022)

Bestman, A.E., Chinyere, J.O., Adebayo, A.A.: Strategic use of information privacy for organizational sustainability. Strat. J. Bus. Change Manage. **9**(3), 517–527 (2022)

Bilderback, S.: Integrating training for organizational sustainability: the application of Sustainable Development Goals globally. Eur. J. Train. Dev. (2023)
Boyd, D.E., Koles, B.: An introduction to the special issue "virtual reality in marketing": definition, theory and practice. J. Bus. Res. **100**, 441–444 (2019)
Bozanic, D., Tešić, D., Puška, A., Štilić, A., Muhsen, Y.R.: Ranking challenges, risks and threats using Fuzzy Inference System. Dec. Mak. Appl. Manage. Eng. **6**(2), 933–947 (2023)
Brydon, M., et al.: Virtual reality as a tool for eliciting empathetic behaviour in carers: an integrative review. J. Med. Imaging Rad. Sci. **52**(3), 466–477 (2021)
Chew, X., Khaw, K. W., Alnoor, A., Ferasso, M., Al Halbusi, H., Muhsen, Y.R.: Circular economy of medical waste: novel intelligent medical waste management framework based on extension linear Diophantine fuzzy FDOSM and neural network approach. Environ. Sci. Poll. Res. 1–27 (2023)
Choi, W., Chung, M.R., Lee, W., Jones, G.J., Svensson, P.G.: A resource-based view of organizational sustainability in sport for development. J. Sport Manage. **1**(aop), 1–11 (2023)
Cook, M., et al.: Challenges and strategies for educational virtual reality. Inf. Technol. Libr. **38**(4), 25–48 (2019)
Dakir, D., Hefniy, H., Zubaidi, Z., Himmah, F.: Early childhood learning mood management through video-based streamlining. Al-Ishlah: J. Pendidikan **14**(1), 157–166 (2022)
Davydov, D.M., Galvez-Sánchez, C.M., Montoro, C.I., de Guevara, C.M.L., Reyes del Paso, G.A.: Personalized behavior management as a replacement for medications for pain control and mood regulation. Sci. Rep. **11**(1), 1–15 (2021)
Demastus, J., Landrum, N.E.: Organizational sustainability schemes align with weak sustainability. Bus. Strategy Environ. (2023)
Dwivedi, P., Chaturvedi, V., Vashist, J.K.: Innovation for organizational sustainability: the role of HR practices and theories. Int. J. Organ. Anal. **31**(3), 759–776 (2023)
Eneizan, B., Mohammed, A.G., Alnoor, A., Alabboodi, A.S., Enaizan, O.: Customer acceptance of mobile marketing in Jordan: an extended UTAUT2 model with trust and risk factors. Int. J. Eng. Bus. Manage. **11**, 1847979019889484 (2019)
Fadhil, S.S., Ismail, R., Alnoor, A.: The influence of soft skills on employability: a case study on technology industry sector in Malaysia. Interdiscip. J. Inf. Knowl. Manag. **16**, 255 (2021)
Farshid, M., Paschen, J., Eriksson, T., Kietzmann, J.: Go boldly!: explore augmented reality (AR), virtual reality (VR), and mixed reality (MR) for business. Bus. Horiz. **61**(5), 657–663 (2018)
Gatea, A.A., Marina, V.: Higher education funding in Iraq in terms of the experience of particular developed countries. Int. J. Adv. Stud. **6**(1), 8–17 (2016)
Gomes, S., et al.: Strategic organizational sustainability in the age of sustainable development goals. Sustainability **15**(13), 10053 (2023)
Hamid, R.A., et al.: How smart is e-tourism? A systematic review of smart tourism recommendation system applying data management. Computer Science Review **39**, 100337 (2021)
Hollebeek, L.D., Clark, M.K., Andreassen, T.W., Sigurdsson, V., Smith, D.: Virtual reality through the customer journey: framework and propositions. J. Retail. Consum. Serv. **55**, 102056 (2020)
Kang, H.J., Shin, J.H., Ponto, K.: How 3D virtual reality stores can shape consumer purchase decisions: the roles of informativeness and playfulness. J. Interact. Mark. **49**, 70–85 (2020)
Khaw, K.W., et al.: Modelling and evaluating trust in mobile commerce: a hybrid three stage Fuzzy Delphi, structural equation modeling, and neural network approach. Int. J. Hum.-Comput. Interact. **38**(16), 1529–1545 (2022)
Kocabulut, Ö., Albayrak, T.: The effects of mood and personality type on service quality perception and customer satisfaction. Int. J. Cult. Tour. Hospit. Res. **13**, 98–112 (2019)
Lahri, V., Shaw, K., Ishizaka, A.: Sustainable supply chain network design problem: using the integrated BWM, TOPSIS, possibilistic programming, and ε-constrained methods. Expert Syst. Appl. **168**, 114373 (2021)

Lopez, F., Brands, F.: Defining firm's competitive sustainability: from fuzzy conceptions to a primer definition and a research agenda. In: 16th Annual International Sustainable Development Research Conference, pp. 1–29 (2010)

Magd, H., Karyamsetty, H.: Organizational Sustainability and TQM in SMEs: a proposed model. Eur. J. Bus. Manage. **13**(4), 88–96 (2021)

Manhal, M., Al-khalidi, A., Hamad, Z.: Strategic network: managerial myopia point of view. Manage. Sci. Lett. **13**(3), 211–218 (2023)

Moriuchi, E., Landers, V.M., Colton, D., Hair, N.: Engagement with chatbots versus augmented reality interactive technology in e-commerce. J. Strateg. Mark. **29**(5), 375–389 (2021)

Muhsen, Y.R., Husin, N.A., Zolkepli, M.B., Manshor, N., Al-Hchaimi, A.A.J.: Evaluation of the routing algorithms for NoC-Based MPSoC: a fuzzy multi-criteria decision-making approach. IEEE Access (2023)

Noland, C.R.: Positive or negative vibes: does mood affect consumer response to controversial advertising? J. Mark. Commun. **27**(8), 897–912 (2021)

Oliari, T.B.P., Stefano, S.R., de Andrade, S.M., da Luz Zaias, L.J.: (2021). Sustainable strategic alignment between people management models of organizational sustainability: insights from Brazil. Revista de Carreiras e Pessoas, 11(1)

Özel, M., Bogueva, D., Marinova, D., Tekiner, I.H.: Climate change knowledge and awareness of nutrition professionals: a case study from Turkey. Sustainability **14**(7), 3774 (2022)

Rao, I., Jha, P.: Gamification in HR for organizational sustainability (2020)

Rizomyliotis, I., Konstantoulaki, K., Kostopoulos, I.: Reassessing the effect of colour on attitude and behavioural intentions in promotional activities: the moderating role of mood and involvement. Australas. Mark. J. **26**(3), 204–215 (2018)

Robinson, M.J., Knobloch-Westerwick, S.: Mood management through selective media use for health and well-being. In: The Routledge Handbook of Media Use and Well-Being, pp. 65–79. Routledge (2016)

Roos, N., Sassen, R., Guenther, E.: Sustainability governance toward an organizational sustainability culture at German higher education institutions. Int. J. Sustain. High. Educ. **24**(3), 553–583 (2023)

Sirakaya, E., Petrick, J., Choi, H.S.: The role of mood on tourism product evaluations. Ann. Tour. Res. **31**(3), 517–539 (2004)

Smilevski, C.: Sustainable leadership and organizational sustainability through organizational change. J. Bus. Parad. **2**(1), 1–33 (2017)

Vágner, V., Bencsik, A.: Organisational knowledge sustainability. In: European Conference on Knowledge Management, vol. 23, no. 2, pp. 1368–1374 (2022)

Violante, M.G., Vezzetti, E., Piazzolla, P.: How to design a virtual reality experience that impacts the consumer engagement: the case of the virtual supermarket. Int. J. Interact. Des. Manuf. (IJIDeM) **13**, 243–262 (2019)

Yin, C.Y., Chang, H.H.: What Is the link between strategic innovation and organizational sustainability? Historical Review and Bibliometric Analytics. Sustainability **14**(11), 6937 (2022)

Yung, R., Khoo-Lattimore, C.: New realities: a systematic literature review on virtual reality and augmented reality in tourism research. Curr. Issue Tour. **22**(17), 2056–2081 (2019)

Author Index

A
Abbas, Sammar 112
Abdel-Hadi, Ilham Nazem 200
Abdullah, Hasan Oudah 112
Aggarwal, Neetima 21
Al Harthi, Aseela 54
Al Harthy, Bushra 54
AL-Abrrow, Hadi 112
Al-Ali, Azhaar 188
Al-Hachim, Hashim Nayef Hashim 69
Al-hashimi, Assmaa Mahdi 188
Alhasnawi, Mushtaq 144
Al-Maatoq, Marwa 159
Almasoodi, Muthana Faaeq 144
Al-Musawi, Hiba Yousif 37
Alshammri, Waleed Saud 94
Al-Sukaini, Adnan Saad Tuama 69
Al-Taie, Mortada Mohsen Taher 200
Arianpoor, Arash 54
Atiyah, Abbas Gatea 83, 144
Atshan, Nadia Atiyah 112
Aubaid Sharif, Bakhtiar 129

B
Baharudin, Siti Mastura 94
Bayram, Gul Erkol 21
Bin Mydin, Al-Amin 175

D
Demir, Beste 1

F
Ferasso, Marcos 37

G
Guven, Selda 1

M
Mohammed, Munaf abdulkadim 159
Mohsin, Abdulridha Nasser 159

P
Prakash, Vijay 21

S
Sahin, Bayram 1
Sharif, Hafiza Saadia 175
Sharma, Neha 21
Shiaa, Hossam Hussein 200

U
Uniyal, Mahesh 21

Y
Younis, Hussain A. 175

Z
Zaidan, Ali Shakir 54
Ziden, Azidah Bt Abu 94

M. Al-Emran et al. (Eds.): IMDC-IST 2024, LNNS 876, p. 211, 2023.
https://doi.org/10.1007/978-3-031-51300-8

Printed in the United States
by Baker & Taylor Publisher Services